THIRD EDITION

# Writing for Technicians

Marva T. Barnett

DELMAR PUBLISHERS INC.®

*Delmar Staff*

Managing Editor: Barbara Christie
Associate Editor: Karen Lavroff
Production Editor: Cynthia Haller

10 9 8 7 6 5 4

**Printed in the United States of America**
**Published simultaneously in Canada**
**by Nelson Canada,**
**A division of The Thomson Corporation**

**Library of Congress Cataloging in Publication Data**

Barnett, Marva T.
  Writing for technicians.

  Includes index.
  1. English language—Rhetoric. 2. English language—
Technical English. 3. Technical writing. 4. Report
writing. 5. English language—Grammar—1950-
I. Title.
PE1475.B34   1987      808'.0666     86-19843
ISBN 0-8273-2833-8 (pbk.)
ISBN 0-8273-2834-6 (teacher's guide)

# PREFACE

*Writing for Technicians* meets the needs for a practical technical communications textbook by both explaining and illustrating fundamentals of technical writing. Its primary purpose is to help technicians and professionals such as engineers, technical communicators, technologists, and social science personnel communicate knowledge of their specialized skills to other interested personnel. At the same time, it emphasizes that technical writers must consider their readers and thus write to express ideas, not to impress these readers.

*Writing for Technicians* is written in the same simple, direct style that it recommends. The text is logically organized. Most sentences are short, the words are carefully chosen, and most information is objective. Therefore, the concepts taught are not concealed behind meaningless phrases, unnecessary words, or involved sentences. Proofreading guides enable students to evaluate their own reports and thus to become increasingly aware of ways to make their writing effective. The specific aids, especially the basic outline, simplify writing techniques. Consequently, anyone who studies the complete text in *Writing for Technicians* and practices writing and evaluating the suggested reports can effectively communicate technical information to others.

Most students in technical writing have studied English grammar, vocabulary, and sentence structure extensively. However, some may benefit from specific suggestions concerning the effective use of grammar in technical writing. For them and also for those wondering about sexist language, Section VI is added. Two other topics discussed in this section — temporary hyphens and the use of numbers — are particularly important in technical writing.

The Instructor's Guide states the general objectives of *Writing for Technicians* and the specific objectives for each unit. The guide is a source of valuable information in its discussion of concepts and guidelines for evaluating written and oral communications. It specifies objectives for the Suggested Activities, suggests guidelines for meeting these objectives, and gives specific answers to specific questions in the Suggested Activities. It also offers suggestions and definitions to help instructors evaluate questions that do not have specific answers.

Students studying *Writing for Technicians* should have learned basic English grammar and the correct use of grammar. Although this textbook was written specifically for two-year colleges and vocational schools, it can be used in the last two years of high school, especially for students preparing to enter industry as tradespeople and technicians. It can also be used in correspondence-study programs or as a self-teaching textbook. People in technical fields in industry can use the text as an aid toward improving the reports they write. The fundamentals of informative writing can be applied to most reports and letters used in any industry, business, or service organization.

*Writing for Technicians* presents and explains the conventions normally followed in preparing technical reports, business correspondence, and other technical writing. However, many of these conventions are not appropriate for published books. For this reason, some of these rules and conventions have not been applied in the composition and layout of this textbook.

The author, Marva T. Barnett, has an extensive background in the teaching of English and technical writing, having taught from the high school to the university level. For fifteen years, she was a technical-writing and practical-English instructor at the University of Utah.

Mrs. Barnett has prepared special courses in English usage and report-writing techniques, which she taught to college students, secretaries, and technical employees of various industrial concerns. In addition, her correspondence study courses in English Grammar Usage and Technical Writing were published for the Institute of Technological Training at the University of Utah. These courses were modified and used by the Utah State Department of Highways, where Mrs. Barnett taught classes in technical writing.

She also taught the correct use of English as part of a communications class in the management program, School of Business, University of Utah. She is coauthor of the textbook *Effective Communications for Public Safety Personnel.* Currently, she is a free-lance consulting editor.

A textbook about such a diverse topic requires the help of many individuals. The contributions of the following are especially appreciated: Jay Smart, Ron Smart, Dr. Edward M. Trujillo, and Mervyn A. Tuttle. John R. Barnett verified much of the technical data.

Note: *Writing for Technicians* was formerly entitled *Elements of Technical Writing.*

## NOTICE TO THE READER

# CONTENTS

## SECTION IV   REPORT SUPPLEMENTS

## SECTION V   WRITING LETTERS

## SECTION VI   EFFECTIVE USE OF LANGUAGE
## IN TECHNICAL WRITING

# Introduction to Technical Writing

# Basic Concepts Related to Technical Writing

## OBJECTIVES

After reading this unit, you should be able to do the following tasks:

- Define technical writing.
- Discuss the importance of technical writing.
- List and describe the eight basic principles of technical writing.

## OVERVIEW

The techniques in *Writing for Technicians* enable a technician to communicate effectively with others. In this unit, technical writing is defined. Objectives, basic principles, and rules used in formal communication are explained. Finally, procedures for organizing and writing technical reports and letters are reviewed.

# TECHNICAL WRITING DEFINED

Writing is divided into two general classes: fiction and nonfiction. *Fiction* tells a story about imaginary characters and situations. Its primary function is to entertain its reader. *Nonfiction* relates only facts. Its primary function is to detail facts and conclusions or interpretations based on facts for the purpose of giving information to its reader.

Technical writing is nonfiction, but further definitions for it vary. At one extreme, it is defined as any information; at the other extreme, it is considered as a highly academic writing style used by scholars in various sciences for scientific papers written only for other scholars.

# DEFINITION FOR TECHNICIANS

Technical writing as discussed in this textbook is somewhere between the two extreme points of view previously mentioned. During the past twenty years, businesspeople, scholars, and teachers have rebelled against reports and letters written only to impress the reader. They want effective communication — a meaningful exchange of information from the writer to the reader. Technicians fulfill their role as writers in this exchange by writing communications that are straightforward, simple, and objective and that also establish rapport with other participants.

# THE IMPORTANCE OF TECHNICAL WRITING

At this point, two questions need to be answered:

1. Why is technical writing important?
2. Why should a student who plans to become a skilled technician or other specialist study a seemingly unrelated subject like writing?

The beginning of civilization and the advent of writing occurred at the same time. Informative writing has undoubtedly been responsible for most of the technological progress made by humanity.

In the early stages of industrial development, around 1750, people needed only simple tools. They lived in isolated groups. Therefore, writing probably was not essential to communication. But even then it was the only permanent and fairly accurate way to pass knowledge on to future generations.

As civilization became more complex, employees often became team members of specialists working toward a major goal such as determining the cause of a plane crash, selecting the best site for a new plant, or protecting the environment. As a result, the need for written communication increased.

Today's employees must not only be specialists in their field but must also be able to write clearly and effectively about their work. Some workers are expected to write progress and justification reports. Others are expected to write detailed formal reports that explain a process or an experiment. Some technicians must research and interpret information. Persons who have repaired faulty equipment may be required to explain the cause of the problem and the procedures involved in correcting it. Often, technicians are sent on field trips and are asked to report to the company about the operation of experimental equipment. Therefore, technicians who can write effectively not only are valuable employees but also enjoy the satisfaction that comes from telling others about their work.

# BASIC PRINCIPLES OF TECHNICAL WRITING

Because people have different reactions to identical experiences, accurate communication is always difficult. Sometimes, the message received by one person is not interpreted to mean at all what the writer had intended. Many communication problems could be avoided, however, if each technical writer recognized the ultimate goal of all informative writing — the meaningful transfer of knowledge from the writer to the reader. Consistent application of the following eight basic principles eliminates most communication problems in informative writing:

1. Understand the reader.
2. Know the purpose of each report.
3. Know the subject matter.
4. Organize the material.
5. Write objectively.
6. Use standard English.
7. Use correct format.
8. Adopt ethical standards.

## Understanding the Reader

Every report should be written for one or more readers who need the information it contains. A writer who understands the reader selects only material that belongs in the report, anticipates the best development of that material, and then expresses ideas in language that cannot be misinterpreted.

Readers of technical reports expect to obtain complete information written clearly and concisely. They usually expect to receive typewritten copy. If handwritten copy is permitted, readers expect it to be legible, to show a clear distinction between lowercase and capital letters, and to be punctuated correctly. They also question the accuracy of facts in a report containing misspelled words.

A report may be read by only one reader or by a variety of readers — managers, supervisors, technicians, new employees, and laypeople. Before technicians prepare reports, they must identify their audience. Because each reader must find the information he or she needs, a report written for many readers includes more detail than a report written for one. It also may require more extensive use of definition and graphic aids. In comparison, a writer may lessen the detail in a report prepared for readers who have general knowledge about the report subject. The writing style, however, does not vary. All reports use standard English, carefully constructed sentences, and logical organization.

## Knowing the Purpose of Each Report

Every report is written to accomplish a specific purpose — sometimes referred to as a goal or an objective. The purpose may be to explain a process, to describe the physical characteristics of some tool or machine, to solve a problem, or to report various kinds of research. (For examples, see Section III, "Writing the Report.")

The importance of clearly and specifically identifying the purpose to be accomplished cannot be overemphasized. The purpose determines content and organization for all written and spoken technical messages and thus establishes the basis for the following processes needed in developing them:

1. Selecting pertinent information and appropriate graphic aids
2. Making the purpose clear to the reader
3. Organizing and presenting text in a way that fulfills the purpose

## Understanding the Subject Matter

Writing job-related reports requires understanding why and how the work was accomplished. For example, a technician who has learned to tune a transmitter turns on the power automatically and rotates various controls clockwise and counterclockwise until a desired frequency is located. Then, the technician adjusts the loading of the transmitter for proper power output. Analysis of how to perform the tuning procedure is no longer necessary. If the technician, however, is explaining the tuning of a transmitter to an uninformed reader, specific information concerning identification, location, use, and operation of controls is necessary. The writer, therefore, must analyze each basic step of the process before writing the report. Only complete explanations including the whys and hows permit the reader to acquire the writer's knowledge.

## Organizing the Material

Organization is important because the reader cannot be expected to understand information that is not logically presented. The report must introduce the reader to the subject, discuss in detail the subject that has been introduced, and then summarize what has been said.

## Writing Objectively

Technical writing should be unbiased and factual. All information is presented from an objective point of view — one that is free from prejudice, preconceived opinions, or emotion. Although conclusions and recommendations are a part of most interpretive and analytical reports, these conclusions and recommendations are not opinions but judgments resulting from a thorough, impartial evaluation of the facts. Using the following rules helps to make a report objective.

- Avoid the use of the personal pronouns *I, we, me, us,* and *you* in formal reports. (This rule does not apply to reports that give instructions or to letters.) The use of these personal pronouns creates a tone of subjectivity and forces the reader to think more of the person involved in the report than of the subject matter being discussed.

- Avoid the use of such adjectives and adverbs as *beautiful, exciting, colorful, slowly,* and *quietly.* Such words create an impression rather than present facts. See Unit 29 "Aids to Conciseness."

- Select words carefully so that the most specific word appropriate for any given situation is used. See Unit 29, "Aids to Conciseness."

- Write simple, straightforward sentences free from generalizations and meaningless expressions. See Unit 29, "Aids to Conciseness," and Unit 30, "Aids to Clarity."

## Using Standard English

All successful craftspeople select quality tools and use them skillfully. A technical communicator's tool is language. Consequently, the quality of language a writer chooses and the skill he or she employs in its use determine the quality of the communication.

Standard English conforms to the sentence structure and vocabulary most commonly understood among English-speaking people. For this reason, it is the only acceptable language for technical reports or letters. Standard English demands the correct use of grammar, punctuation, spelling, and temporary hyphens (Unit 24). It also demands uncluttered sentence structure

(Units 25, 29, and 30), words that say what they mean (Unit 28), and the effective use of numbers (Unit 26).

Standard English rejects language characteristic of any special group or locality, including generalities, slang, colloquialisms, archaic expressions, and cliches (Units 18 and 30). Finally, it rejects any individually coined words including the *-ize* and *wise* words such as *perceptualize, moneywise,* or *sciencewise* that have emerged in recent years.

## Using Correct Format

Format, the arrangement of headings, text, and supplements, is responsible for the general appearance of a report. Because format is the first thing a reader sees, it is as important as language, style, and report text.

Some readers, if they have authority to do so, justifiably refuse to read reports that look unattractive. If they do read them, an unfavorable first impression can influence their reaction to the entire report. They are inclined to criticize minor errors, to question the validity of information, and to reject any included proposals or recommendations. Consequently, writers can spend hours preparing a complete, informative report and still risk its rejection, regardless of its content, when they fail to create an attractive format.

## Adopting Ethical Standards

Good technical writers are ethical and honest. They objectively present all the facts when such facts are available and justify any information that is not available. They evaluate facts objectively without bias and without consideration of personal gain. They base their proposals and recommendations upon information and judgment, not upon preconceived opinions. In addition, they carefully document subject matter that is not their own and acknowledge others who have helped them prepare their reports.

# SUMMARY

1. Technical writing is an essential part of today's industrial, business, and science society.

2. Technical communication involves sending and receiving informative messages such as reports, memorandums, and letters.

3. A technical writer is obligated to write facts and interpretations of facts clearly, concisely, and completely enough for an uninformed reader to accurately understand them.

4. An attitude of cooperation must exist between any writer and reader involved in technical communication.

5. An effective technical writer understands and applies eight basic principles that are essential in all informative communication.

UNIT

# SUGGESTED ACTIVITIES

A. Answer the following questions:

1. What does *technical communications* mean to you?
2. The term *effective communication* is used frequently in this textbook. What meaning does the term have for you?
3. As a person trained in a special trade or technical field, why should you be expected to write effectively? Justify your answer; but after evaluating the principles discussed in this unit, base your answer upon your own convictions, not just upon the ideas expressed in this unit.

B. Be prepared to discuss in class or as a written assignment the eight basic principles of effective technical writing. (These principles should be thoroughly understood and remembered so that you can apply them to all your written communications.)

C. Do you agree that applying these basic principles is necessary to communicate effectively? Justify your answer.

D. If possible, obtain a technical communication for use in classroom discussion. (Companies or social service organizations often permit students to use one from their files.) Determine whether the communication is effective. Also, determine whether any basic principles discussed in this unit have been ignored.

# The Writer's Attitude

After reading this unit, you should be able to do the following tasks:

- Describe the role of personal attitudes in written reports.
- List and describe the seven personality traits that affect writing.

## OVERVIEW

Men and women engaged in trades or technology sometimes resent the time and concentration needed to prepare good reports and letters. However, the employees who are serious about performance know that success depends largely upon effectively reporting the results of their work. They also realize that they alone understand the work well enough to report it accurately and completely.

This textbook is designed to help these employees recognize the value of effectively communicating information, prepare messages in a minimum amount of time, and avoid the frustration that unskilled writers and speakers usually experience. Equally important, this unit encourages writers to recognize that attitude contributes significantly to effective communication and develop techniques that reflect a confident, positive, and cooperative attitude.

# THE ROLE OF PERSONAL ATTITUDES IN COMMUNICATIONS

Reports and letters sent from one person to another do more than transmit information. They also reflect some of the character traits of the sender, even though the content is expressed objectively. In other words, a person's attitude is often evident in the style and content of the message. For example, a dictatorial message may convey a writer's attitude of superiority to the reader. A message that has an unattractive format suggests that the writer lacks pride in his or her work and has little, if any, concern for the reader. A poorly organized message suggests that the writer did not or cannot think clearly. One that is incomplete indicates impatience, negligence, or both. Anger is easily recognized, especially in letters. A person listening to a speaker (Unit 15) can easily tell whether the speaker cares about the audience or is merely talking.

# PERSONALITY TRAITS FOR EFFECTIVE COMMUNICATION

Because attitude is so apparent in communications, anyone engaged in technical communications can benefit from developing seven personality traits that contribute significantly to effective technical writing:

1. Self-confidence
2. Cooperation
3. Judgment
4. Empathy
5. Patience
6. Integrity
7. Objectivity

These traits are relative in contrast to absolute. In other words, they vary in degree. For example, probably no one can always be completely confident, completely cooperative, and so forth. The degree to which these traits are acquired usually depends upon a person's knowledge; experience; and analysis of self, other people, and situations.

## Self-Confidence

Self-confidence is the most valuable personal trait any person can possess, partly because it permits the other six traits to be developed. A self-confident attitude is not easily or quickly acquired. However, the personal and professional rewards it brings more than pay for the time, energy, and effort required to develop it.

Self-confidence, not to be confused with overconfidence or arrogance, is a firm belief in one's self and in one's own abilities. Individuals striving to become self-confident periodically ask themselves what they are capable of doing; what they cannot do; and how they can improve themselves, their job skills, and their relations with others.

## Cooperation

Cooperation is the act of working harmoniously with others to achieve a common goal. In technical communications, usually the common goal is to exchange information in a way that allows all participants to reach a common understanding that results in productive action.

Because cooperation implies a willingness to consider another person's reactions and possible reasons for the reactions, it is closely allied to self-confidence. Only a confident person can display an attitude of cooperation without feeling or acting submissive or without fearing that someone will take advantage of this attitude.

## Judgment

Judgment is formally defined as the ability to observe situations carefully enough to understand them, to make logical decisions concerning them, and then to take positive action when it is needed. It is also defined as good sense.

A few people seem to be born with the ability to make the right decision and to act upon that decision. Usually, however, this quality is developed slowly. Many rules serve as guidelines for effective technical writing. Nevertheless, messages cannot always be fit into standard patterns and therefore cannot always be prepared according to these standard rules. A good writer develops the ability to recognize when or when not to deviate from the rules. Therefore, judgment is the personal characteristic that allows a writer to avoid stereotyped communications that often become dull and ineffective. (Further references to using judgment are made throughout this textbook.)

## Empathy

Empathy (not to be confused with sympathy) simply means having the ability to imagine how another person feels about something. In technical communications empathetic writers ask themselves the following questions: "How will the recipients respond to this message? Will they understand it? Will they react favorably? Will the way I've presented it bring the desired results?" Thus, technical writers cannot be concerned only with what they want to say and the way they want to say it. Unless they recognize the needs of their correspondents and strive to fulfill those needs, effective communication is not likely to develop.

## Patience

In communications, patience means the ability to work steadily to achieve a goal. The goal is to present data clearly, concisely, and accurately in a style and format that build understanding and cooperation between the sender and the receiver. Hastily written reports and letters are usually sketchy and unorganized. They reflect little consideration for the reader or for the information being reported. In turn, readers are inclined to ignore them or consider them unimportant. As a consequence, impatience often results in lost time and effort for both the writer and the reader. In contrast, patience allows a concerned writer to develop four important communication skills:

1. Organize a report, memorandum, or letter logically.
2. Include all necessary, specific details.
3. Eliminate unnecessary ideas.
4. Select words and well-organized sentences that prevent misinterpretation.

## Integrity

Integrity is the quality of being sincere, honest, and candid. Though sometimes difficult to acquire and maintain, integrity is a personal trait that ultimately frees a person from being identified with any kind of deception or shallowness.

Integrity in communications permits a person to be honest in reporting exactly all required information without concern for consequences. Unless facts are reported honestly, meaningful communication cannot be achieved.

A writer who combines integrity with empathy develops the ability to report only facts or opinions based on facts but to report them tactfully.

## Objectivity

Objective writing is defined in Unit 1 as writing that is unbiased and factual. This kind of writing is possible only when the individuals involved in a communication maintain an attitude free from prejudice or prejudgment. They do not allow personal interest or personal feelings to influence technical communication.

Objectivity is not isolated from other attitudes such as self-confidence and empathy, sometimes classed as emotions. A self-confident person is not easily influenced by politics or unsupported opinions from others. Therefore, he or she is free to be objective. Empathy builds the understanding and cooperation needed to collect and verify data received from others.

# SUMMARY

1. Personal attitudes influence all technical communication positively or negatively, depending upon the attitude the writer adopts.

2. A strong relationship exists between the basic principles of technical writing (Unit 1) and a writer's attitude.

   a. Self-confidence contributes to objectivity and the adoption of ethical standards.
   b. Cooperation contributes to understanding the reader and knowing subject matter.
   c. Judgment contributes to organizing material and adopting ethical standards.
   d. Empathy contributes to understanding the reader.
   e. Patience contributes to effective organization of data, use of standard English, and correct format.

3. A writer is not likely to perfect the attitudes discussed in Unit 2. They are qualities that continually change.

4. As mentioned frequently in Unit 2, a strong interplay exists among attitudes, one attitude contributing to another.

5. Self-confidence is probably the most important attitude; without it, a person has difficulty developing the others.

---

UNIT

# SUGGESTED ACTIVITIES

A. Answer the following questions:

   1. Explain the difference between self-confidence and arrogance. (Use a standard dictionary to help with your answer.)
   2. Why would arrogance be detrimental in technical communications?
   3. List at least three characteristics that indicate a person is self-confident.
   4. Do you believe self-confidence helps an employee communicate technical data effectively? Justify your answer.

B. In a classroom discussion be prepared to explain how empathy helps you develop a cooperative attitude toward others.

C. Is it fair to ask a technician, usually a physically active person, to be patient enough to write meaningful reports as part of his or her job? Be prepared to justify your answer.

D. You may have known people who, in your opinion, lack integrity but are considered successful. Do you believe that integrity as a personal trait is important? Be prepared to justify your opinion.

E. Bring to class a report or business letter. Can the writer's attitude be determined in the tone, format, and content of the communication?

SECTION

# SUGGESTED SUPPLEMENTARY READING

Caernarven-Smith, Patricia. *Audience Analysis and Response.* Dedham, Mass.: Firman Technical Publications, Inc., 1985. The preface gives some interesting insights into audience response and communicator-audience interaction. Pages 129 to 167 offer many practical aids for building effective communication between the writer or speaker and the audience.

Damerst, William. *Clear Technical Reports.* New York: Harcourt Brace Jovanovich, Inc., 1972. Several major obstacles to a reader's ability to understand a written message are discussed on pages 11 to 18.

Houp, Kenneth W., and Thomas E. Pearsall. *Reporting Technical Information.* New York: Macmillan Publishing Company, Inc., 1984. On page 4, three elements needed for reporting technical information and the importance of knowing the reader are discussed.

Lannon, John M. *Technical Writing.* Boston: Little, Brown and Company, 1985. Technical writing is defined on page 3. Six features of technical writing are listed on pages 5 and 6. Relating to the audience (reader) is discussed on pages 9 to 11.

Markel, Michael H. *Technical Writing: Situations and Strategies.* New York: St. Martin's Press, 1984. Characteristics of effective technical writing are discussed on pages 5 to 8. Different kinds of audiences (readers) are described on pages 13 to 19.

Sherman, Theodore A., and Simon S. Johnson. *Modern Technical Writing.* Englewood Cliffs, N.J.: Prentice-Hall, Inc., 1983. "Introduction," pages 3 to 8, tells why the ability to write effective technical reports is important to technical employees.

The following books are suggested as sources for developing the personal traits introduced in Unit 2, "The Writer's Attitude." The entire book is useful unless specific pages are given.

Brandon, Nathaniel. *Honoring the Self.* Section I: "The Dynamics of Self-Esteem." Englewood Cliffs, N.J.: Prentice-Hall, Inc., 1983.

Branden, Nathaniel. *The Psychology of Self-Esteem*, pages 109 to 139. New York: Bantam Books, Inc., 1973.

Cooper, Alfred M. *How to Supervise People.* 4th ed. Chapter 5, pages 74 to 89. New York: McGraw-Hill Book Company, 1963.

Dale, Ernest. *Management: Theory and Practice.* "The Theory and Practice of Decision Making," page 547. New York: McGraw-Hill Book Company, 1969.

Dyer, Dr. Wayne W. *Pulling Your Own Strings.* New York: Funk & Wagnalls, Inc., 1978.

Dyer, Dr. Wayne W. *Your Erroneous Zones.* New York: Funk & Wagnalls, Inc., 1976.

Glasser, Dr. William. *Reality Therapy,* pages 1 to 80. New York: Harper & Row, Publishers, Inc., 1975.

Lair, Jess, Ph.D. *I Ain't Much, Baby — but I'm All I've Got.* Greenwich, Conn.: Fawcett Publications, Inc., 1974.

Maltz, Dr. Maxwell. *The Search for Self Respect.* Chapter 13: "Integrity as a Way of Life." New York: Bantam Books, 1976.

Trimble, John R. *Writing with Style.* Englewood Cliffs, N.J.: Prentice-Hall, Inc., 1975. The personal qualities of confidence (page 3), positive attitude (page 5), objectivity, empathy, and courtesy (pages 16 to 19) are briefly discussed. On pages 19 to 23, five ways to serve a reader's needs are listed and discussed.

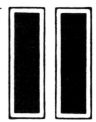

# Planning
# The Report

# Definition in Reports

## OBJECTIVES

After reading this unit, you should be able to do the following tasks:

- Describe and write an informal definition.
- Describe and write a formal definition.
- Describe and write an expanded definition.
- List and describe the ten techniques that can be used in an expanded definition.
- Determine where a definition should be placed in a report.

## OVERVIEW

Definitions are used in reports to clarify technical terms that may be unfamiliar to one or more readers, not to clarify words chosen for general discussion. As explained in Units 28 and 30, words in general text must be precise enough to clearly express the intended meaning.

This unit discusses definition as an explanation that gives the exact meaning of a word or a group of words. Some technical terms can be defined in only a few words or in one or two sentences. For these terms, a writer may use an informal or a formal definition. Other terms must be defined more fully. For these, a writer uses an expanded definition, one that includes a formal definition and additional information. Only the person writing a specific report can determine the kind of definition needed to make the reader understand the information being presented.

# INFORMAL DEFINITION

An informal definition is a familiar word or phrase that immediately follows the technical term being defined.

The example illustrates a one-word informal definition. The writer has determined that a more detailed definition is unnecessary.

---

**EXAMPLE**

**The sample contained cretaceous (chalklike) rock.**

---

In the next example, the informal definition is written as a phrase, which may be enclosed in parentheses or set off by commas from the rest of the sentence.

---

**EXAMPLE**

**A photometric analyzer (an instrument that measures light intensity) uses a mercury-vapor lamp as a source.**

---

Although informal definitions aid understanding, they can, if overused, interrupt important ideas. Therefore, if more than two unfamiliar technical terms in a three- or four-page report require definition, all definitions should be written formally and placed in the introduction to the report or in a glossary.

# FORMAL DEFINITION

A formal definition is written as one or two complete sentences and consists of three parts:

1. The word to be defined
2. A general class to which the word belongs
3. The characteristics of the word being defined that distinguish that word from other members of the general class

---

**EXAMPLE**

| Defined Word | General Class | Distinguishing Characteristics |
|---|---|---|
| pulsimeter | an instrument | measures the alternate expansion and recoil of an artery caused by contractions of the heart |

---

The information in this example becomes a formal definition only after it is written as a complete sentence.

**EXAMPLE**

**A pulsimeter is an instrument that measures the alternate expansion and recoil of an artery caused by contractions of the heart.**

This formal definition is expressed in one complete sentence. Some formal definitions are more clearly expressed when two sentences are used.

**EXAMPLE**

**A hammer is a hand tool. It has a solid head set crosswise on a handle and is used primarily for pounding.**

Writing a good formal definition requires that the writer observe the following six guidelines.

- The class to which the word belongs must be expressed in specific terms, as illustrated by *hand tool* in the preceding example.

- The distinguishing characteristics must be detailed enough to distinguish the word being defined from other items included in the general class. If, for example, the description "It has a solid head set crosswise on a handle and is used primarily for pounding" does not distinguish a hammer from other hand tools, the definition has failed.

- Another form of a word being defined is not to be used as part of the definition. The following sentence illustrates how ineffective a definition can be when this guideline is ignored.

**EXAMPLE**

**Cancellation is the result of canceling a number.**

An exception to this rule is made when two words naming an item are being defined but one of the two words is familiar to the reader. For example, a gear wheel may be defined as a wheel that has a rim notched into teeth that mesh with another wheel to receive or transmit motion. A reader understands the meaning of *wheel;* therefore, *wheel* can become part of the definition.

- Words less familiar than the word being defined are not used as part of the definition.

**EXAMPLE**

**Gene mutation is a change caused by fundamental intramolecular reorganization of a gene.**

This definition has accomplished nothing. In the attempt to define *gene mutation,* the writer has used the unfamiliar term *intramolecular* and has repeated the word *gene,* which may not be understood by the reader.

- Some technical terms have more than one meaning. For example, a cylinder means one thing to an engineer and something else to a newspaper printer. A qualifying statement, therefore, is included in a definition written for one specific purpose.

**EXAMPLE**

**The word *cylinder,* as used in this report, means a chamber in which force is exerted on the piston of a reciprocating engine.**

The use of the qualifying words "as used in this report" prevents misinterpretation and keeps a reader from questioning whether the definition is complete and correct.

- The purpose of a definition is to make the reader understand the meaning of an unfamiliar word. Any definition that does not assist the reader should be omitted from the report.

# EXPANDED DEFINITION

Sometimes, the meaning of an entire report depends upon the reader's understanding of a technical item that cannot be described in a few words or a sentence. A writer then uses an expanded definition, a detailed explanation that follows the formal definition. For example, a report that explains how the height of a standing tree is measured should probably include an expanded definition of a clinometer, the instrument used to make the measurement.

As shown in Figure 3–1, the expanded definition of the clinometer begins with a formal definition. Then, the writer uses other kinds of explanations until the reader can visualize a clinometer, understand the principle of operation, and determine how the clinometer is used.

The expanded definition in Figure 3–1 begins with a formal definition; then it explains the physical characteristics, the basic principle, and the use of a clinometer.

A clinometer is a hand-held instrument that measures angles of elevation. The clinometer used to measure a standing tree is 3 inches long, 2 inches wide, and 3/5 of an inch deep. It is slightly rounded at each end, as shown in Figure 1.

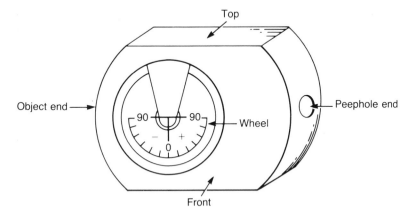

FIGURE 1.  A Clinometer

Inside the clinometer is a wheel sealed in alcohol. The wheel is balanced so that the bottom of the wheel marked 0 degrees always points to the center of the earth. On the face of the wheel are markings that measure degrees from minus 90 degrees through 0 degrees to plus 90 degrees. These markings are visible through a plastic window located on the front side of the clinometer *and* through a peephole located at one end.

The person using the clinometer stands 100 feet from the tree, holds the peephole of the clinometer to one eye, and sights the top of the tree. At the same time, by means of an internal optical device, the person reads the degrees of elevation to the top of the tree. On the back side of the clinometer is a table that converts degrees of elevation into height of the tree for the horizontal distance of 100 feet.

FIGURE 3–1  An example of an expanded definition

One or more of the following ten techniques may be used in an expanded definition. The amount of information needed to clarify the technical term being defined determines the definition length.

## Analysis

In analysis, a device is separated into its major parts, or a process is divided into the major steps needed to complete the process.

**EXAMPLE**

**All digital computers have five basic parts:**
1. **An input device**
2. **An output device**
3. **A memory device**
4. **A logic or an arithmetic section**
5. **A control section**

## Basic Principle

A basic principle is a fundamental rule or a scientific rule that permits something to operate in a particular way. For example, a thermometer is constructed on the principle that mercury expands when heated. Application of this principle is the reason mercury rises in a confined cylinder.

## Classification

All items belonging to a general class may be listed as part of an expanded definition. A detailed discussion of classification and its relationship to expanded definition is given in Unit 4.

## Comparison and Contrast

An unfamiliar device or process may be effectively defined when similarities and differences between it and a familiar device or process are explained.

In the following example, the similarities and differences between a pi tape and the more familiar tape measure are explained.

**EXAMPLE**

The pi tape is similar in appearance to the common tape measure used by carpenters and other craftspeople. Like a tape measure, the pi tape is placed around the circumference of a spherical object to determine diameter.
In contrast, the pi divisions on the pi tape give a direct reading of diameter instead of circumference.

## Etymology

Sometimes, understanding the origin of a word or the changes in meaning that have occurred to give a word its present meaning is a useful part of a definition. For example, *component* is derived from the Latin word *componere*, meaning "to put together." A *component* is a constituent part or an ingredient.

Knowing the etymology of *component* helps a reader understand the statement "A radio is made up of many components." The etymology of a word is usually found in standard dictionaries as part of the definition.

## Graphic Aids

A most effective technique often used in an expanded definition is to show a picture of a device or process combined with verbal references to the picture. The picture must be given a figure number and a name; each part of the picture must also be identified. A statement similar to "Figure 2 illustrates meshed gears" introduces the picture. Any further explanation is written after the picture has been presented. See Figure 3-2.

Figure 2 illustrates meshed gears.  As indicated by the arrow, meshing occurs each time a tooth of one gear interlocks with a tooth of another gear.

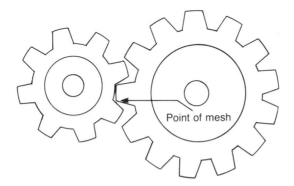

Point of mesh

FIGURE 2.  Meshed Gears

FIGURE 3-2 An example of a graphic aid technique

## Location

Telling a reader where the object being defined is located often aids understanding. For example, knowing that manganese nodules are found in oceans will help the reader understand problems associated with mining them.

## Negative Statement

A negative statement tells what a device or process is not.

| EXAMPLE | **Digital computers should not be confused with another class of computers known as analog computers.** |
|---|---|

This example tells what a digital computer is not, but it does not tell what a digital computer is. Therefore, the definition has been started but not completed. As a single method of definition, the negative statement has little value; but when used with other techniques, it can be effective.

## Physical Description

Physical description is a description of any characteristic that is part of the material makeup of a device — size, color, texture, and shape. Combined with a graphic aid, as illustrated in the definition of the clinometer, physical description is an important kind of definition.

## Specific Example

Specific example explains the circumstances under which a device or process is used. A specific example that can be related to the experience of the reader is an effective method of expanding a definition.

| EXAMPLE | **A variable-speed drive is used on power drill presses to vary the drilling speed.** |
|---|---|

The reader who is familiar with drill presses can understand this example and, therefore, can more easily recognize the use of a variable-speed drive in unfamiliar devices.

Each technique used in an expanded definition evolves logically from the preceding one so that the completed definition is unified and coherent. This progression can be seen in the definition of the clinometer. Any obvious shift from one technique of expanded definition to another can interrupt the development of meaning.

# DEFINITIONS IN REPORTS

Definitions may be placed in one of three parts of a report:

1. In the introduction to the report
2. In the text
3. In a glossary

A *glossary* is a collection of terms limited to a special area of knowledge.

Informal definitions are placed either in the introduction to a report or in the text — wherever the terms being informally defined are first used.

Most formal definitions are placed in a special definition section in the introduction to the report. In this section, any technical terms that are used anywhere in the report and that require a formal definition are introduced and defined.

**EXAMPLE**

**The following definitions clarify the technical terms used in this report.**

After this introductory statement, each technical term is written as a paragraph heading. The definition, written in one or two complete sentences, follows the paragraph heading, as illustrated in Figure 3-3. Also see Figure 7-4.

---

**Annealing.** Annealing is a process of heating and slowly cooling glass or metal to prevent brittleness.

**Elasticity.** Elasticity is the property a body possesses to assume its original shape after a deforming load has been removed.

**Groove Weld.** A groove weld is a weld made by the deposit of filler metal into a groove between two metals that are to be joined.

---

FIGURE 3-3  An example of formal definitions of technical terms

Sometimes, a writer cannot be sure whether a reader will need definitions and, therefore, hesitates to include them in the report. Also, lengthy reports containing highly technical data may require numerous definitions. Under these circumstances, formal definitions can be placed in a glossary. The reader is informed in the introductory section reserved for definitions that a glossary has been written.

Placing formal definitions in footnotes is discouraged because footnotes distract the reader's attention from the topic being discussed. Any definition that the writer considers placing in a footnote can probably be more effectively placed in the introduction to the report.

Expanded definitions are usually placed in the special section for definitions in the introduction to the report. However, sometimes an expanded definition must be placed in the text because the definition relates to one major topic and will be more effective if it is associated with that topic. Then, the definition may immediately follow the sentence that introduces the topic.

**EXAMPLE**   **To extract the valuable minerals from minerals that have no value, mining companies use one of two processes: (1) leaching or (2) flotation.**

After this introductory sentence, an expanded definition for each of the two processes, leaching and flotation, may be given. Frequent use of expanded definition within the text, however, can create confusion.

# SUMMARY

1. Definitions in technical communications are used only for technical terms, not for words used in writing the general text.

2. Three kinds of definitions useful in technical communication are (1) informal, (2) formal, and (3) expanded. The choice depends upon the amount of information needed to clarify a technical term for all readers.

3. Definitions are preferably placed in the report introduction (Unit 6) or in a glossary. The number of definitions used and the number of readers likely to need them help determine location. Excessive definitions (more than four) in the introduction can confuse most readers.

4. Except when an occasional expanded definition is needed, all definitions in the report introduction or glossary are formal definitions. Expanded definitions are placed in the report introduction or body of the report, never in a glossary.

5. Informal definitions, used sparingly, follow in parentheses the word being defined.

6. A report should rarely require more than one expanded definition. Only a subject important to the entire report should require extensive explanation.

7. A writer, after evaluating his or her audience and report subject, uses judgment to determine the number and kind of definitions needed.

8. A statement in the report introduction always introduces formal and expanded definitions or tells where they can be found.

UNIT

## SUGGESTED ACTIVITIES

A. Write an informal definition for each of the following terms. The familiar words *resistance*, *weld*, and *matter* may be used as part of the definition. References may be consulted, but the definitions must be original.

1. Component
2. Exothermic
3. Heat loss
4. Horsepower
5. Integer
6. Ion
7. Kinetic
8. Microseism

9. Nonlinear resistance
10. Organic matter
11. Peen (noun)
12. Tack weld
13. Thermal conductivity
14. Software
15. Viscosity

B. Using the rules for formal definition, rewrite the following definitions. If the meanings given are not clear, use a dictionary, textbook, or encyclopedia to help clarify the term being defined.

1. An amplifier is something in electricity and radio that increases the strength of electric impulses.

2. Generator — a machine for changing mechanical energy into electric energy.

3. Conductance is the reciprocal of resistance.

4. Internal gears have their teeth formed inside the root circle.

5. Input is when punched cards, punched paper tape, and magnetic tape are used to enter information into a computer.

6. A tape controller is used to synchronize the speed of the tape unit with the speed of the computer.

7. The hydraulic pump (operated by the force and movement of water).

8. Annealing is softening metal.

9. Coulter counter. It samples, dilutes, counts, analyzes, and reports out red and white blood cells.

10. Iatrogenic disease arises from diagnostic and therapeutic procedures used to treat another disease.

C. Write an expanded definition for a technical term that requires more explanation than can be given in a formal definition. In the left-hand margin, write the techniques of definition you have used. (See the following examples.)

**EXAMPLES**

| | |
|---|---|
| *Formal Definition* | A pi tape is a measuring device used to determine the true diameter of a spherical object to the nearest 0.001 of an inch. |
| *Derivation* | The name *pi tape* is derived from the Greek letter pi, which designates the ratio of the circumference of an object to *the diameter*. |
| *Physical Description* | A pi tape is a narrow band of steel 1/4 of an inch wide and 1/16 of an inch thick. The length of the tape varies and is determined by the range of diameters to be measured. A rolled handle attached to the leading end of the tape aids in tightening the tape against the object being measured. |
| *Physical Description and Basic Principle* | Three inches from the handle is a 2-inch-long vernier scale that gives a diameter reading of 0.001-of-an-inch accuracy. The rest of the pi tape is marked off in measurements of diameter that eliminate a mathematical conversion from the measurement of circumference. |

This example is not a guide for length or techniques to be used by the students completing this activity. The objective in this assignment is to write the kind and amount of information that helps the reader understand the term being defined.

# Classification in Reports

After reading this unit, you should be able to do the following tasks:

- Define and describe the three kinds of classification.
- Use the three kinds of classification in reports.
- Distinguish between partition and classification.

## OVERVIEW

A classification is simply a *systematic* arrangement of things that belong to one general class. A general class contains two or more items that have one or more similar characteristics. For example, *tall*, *medium*, and *short* have the common characteristic of height. Thus, when these or similar items are listed in some logical arrangement under a statement, a heading, or a title that identifies the general class, a classification exists.

This unit explains that classifications can be simple, including only general items, or increasingly specific. They can be presented as a list of related items, as an outline, as a graphic aid, or as expository text.

# KINDS OF CLASSIFICATION

To be meaningful in technical writing, classifications must be divided into three different kinds:

— Unlimited classification
— Limited classification
— Formal classification

## Unlimited Classification

An unlimited classification is usually a list of items systematically arranged under a general heading that is not restricted by a qualifying word or phrase. For example, *minerals* or *kinds of minerals* is an unrestricted general heading. When all major classes of minerals are arranged alphabetically or chronologically under the general heading, a simple unlimited classification results. If one or more of the listed classes is then subdivided into more specific minerals, the unlimited classification can become long and involved.

On the other hand, an unlimited classification can be short. For example, computers can be divided into just two major classes: digital and analog. Similarly, blood can be divided into only four major classes: A, B, O, and AB.

The following example shows an unlimited classification similar to many that technical writers can use to present data clearly and concisely.

**EXAMPLE**

**SOURCES OF HEAT**

**Body**

**Electricity**

**Friction**

**Fuel**

**Hot springs**

**Sun**

Six general types of heat are listed in alphabetical order under a heading that identifies a class to which they belong. If each listed item were used as a heading, and it could be, a more specific but still unlimited classification would result. Also, if sources for one or more of the listed items, such as coal, coke, gas, and so forth, were listed as sources for fuel, an unlimited classification written as an outline would result. An outline format is shown in "Types of Metamorphic Rock" (p. 38).

## Limited Classification

A limited classification is one in which the general class is limited by a qualifying word or phrase. Therefore, the unlimited classification of *minerals* becomes limited when the minerals included are restricted to a particular group of minerals. For example, a classification of minerals found only in compounds is a limited classification. A list of minerals used in a particular industry is also limited classification. Only one limitation may be used for any one classification. The limitation is clearly stated in a heading placed above the classification, a lead-in sentence, or a short heading followed by a lead-in sentence.

EXAMPLE

**A CLASSIFICATION OF MINERALS USED IN MANUFACTURING**

Under this heading only the minerals used in manufacturing are included, but all minerals used in manufacturing must be included. This list, which would still be rather general, can be made more specific if one type of manufacturing is specified.

EXAMPLE

**A CLASSIFICATION OF MINERALS USED IN THE MANUFACTURE OF AIRPLANES**

After a classification has been limited in a way that is meaningful to a reader and appropriate for the subject being discussed, it is a valuable aid to clarity in technical reports.

The following example shows a lead-in sentence introducing a limited classification.

EXAMPLE

**The following chemicals were used in the experiment:**

A heading followed by a lead-in sentence again limits the classification.

EXAMPLE

**LIST OF EQUIPMENT**

**The specific tools and test equipment needed for the tune-up are listed in the order in which they are used:**

## Formal Classification

A formal classification is one in which the items in a limited general class are arranged according to some designated basis. The *basis* is a word or phrase that permits classified items to be compared with one another in a specific way. For example, if the classification of minerals used in the manufacture of airplanes is arranged according to the specific use for each, the specific use is the basis for one formal classification. Similarly, types of lumber used for constructing buildings can be classified on the basis of durability because all lumber has some durability, but one type of lumber is more durable than another. (For another example, see Figure 4–1.)

# CLASSIFICATION IN REPORTS

Classification helps to define, clarify, or develop subject matter that can be more easily interpreted when it is divided into categories. A technician may need to include a classification of soils in a report concerning a proposed highway. A designer explaining the development of new equipment may classify materials considered for use in the equipment. Similarly, a limited classification is used to list materials or equipment discussed in a process report. A person writing about potential locations for new mines probably classifies ores found in a specific area. As indicated by these examples, classification is a basic part of all technology and is, therefore, often an effective tool in technical report writing. Judgment and consideration for the reader help a writer decide when classification should be used and which type and format will be most meaningful.

## Using Classification

The following guidelines are given to help a writer use all kinds of classifications effectively in technical reports.

- Classification is not to be confused with partition. *Partition* means to divide one thing into its parts. Partition is used as part of many descriptive reports. For example, a screw-holding screwdriver has three parts: a handle, a shank, and a slide. *Classification* means to divide a group of things into more specific items. A list of all screwdrivers, if arranged systematically, is a classification.

- A classification used in technical writing includes only facts that can be verified. Vague words such as *strong, quite strong, weak,* and similar adjectives are not acceptable descriptions unless a scale, by which the adjectives can be interpreted, has been given to the reader.

- Symbols and abbreviations are used only when space is limited. Even then, they must be meaningful to the reader.

- A classification must be complete. All items belonging to a designated heading or title must be included in the classification.

- A classification presented as a list, an outline, or a table is an integral part of the text. It is introduced, presented, and discussed.

## Using Unlimited Classification

If an unlimited classification — as in classes of chemicals, green vegetables, and similar lengthy topics — exceeds five, the classification becomes more distracting than useful. Any classification of this kind needed as a reference should be placed in an appendix.

If the list of major classes does not exceed five, as in classes of blood, coal, or computers, an unlimited classification can help organize and simplify report data. For example, an expanded definition can include a brief discussion of computers and the two major classes of computers: digital and analog.

**EXAMPLE**   A computer is a mechanical or electronic device that performs mathematical problems. Computers are divided into two classes: digital and analog.

Additional text follows if the purpose for using the classification requires it. A writer may prefer to use a list or outline or combine a list or outline with text for longer or more detailed classifications.

## Using Limited Classification

A limited classification in a report may be systematically arranged into a list, an outline, or a sentence or paragraph. As a list, the classified material is usually preceded by a lead-in sentence in which the limited class is identified.

Unless a writer has a specific reason for using another arrangement, the list is arranged alphabetically and each item is preceded by an arabic numeral.

**EXAMPLE**

**Gas welding is divided into four specific types:**

1. Air-acetylene welding
2. Oxyacetylene welding
3. Oxyhydrogen welding
4. Pressure gas welding

Some limited classifications are written as an outline when items under the general class are subdivided into more specific parts.

**EXAMPLE**

**TYPES OF METAMORPHIC ROCK**

I. Foliated rock
   A. Gneiss
   B. Phyllite
   C. Schist
   D. Slate
II. Nonfoliated rock
   A. Marble
   B. Metaconglomerate
   C. Quartzite

An outline classification begins with a heading rather than a lead-in sentence. The major items under the heading, two spaces under the centered title, begin with a roman numeral. Capital letters listed in alphabetical order introduce each item in the second-level list. The first letter of all listed items is capitalized. In a more detailed classification, arabic numerals and lower-case letters precede third- and fourth-level lists; however, three- and four-level lists tend to confuse a reader. Limiting a heading to include only pertinent information prevents the need for detailed outlines.

A limited classification, especially one that is part of an expanded definition or an analysis, may be written as one or more sentences or a paragraph.

**EXAMPLE**

**Sodium, lithium, and potassium are the three metals always used in alloys. They are silver-white metallic elements that are never found in a pure state in nature. All three metals are lighter than water, but lithium is lighter than either potassium or sodium.**

The topic sentence in this example limits the classification to metals used in alloys. The paragraph explains only the similarities of the three metals. If additional paragraphs follow to explain characteristics that distinguish the three metals from one another, the classification is formal.

## Using Formal Classification

Formal classification, an essential part of technical writing, is the basis of comparative analysis, a type of report used extensively in business, industry, and social service organizations. In reports, formal classifications are presented in the form of tables and text, text, or text and illustration.

In a table (Figure 4-1), a formal classification is arranged as two columns of information following a formal title for the classification and a subtitle for each column. Other tables are discussed in Unit 5, "Graphic Aids."

The following guidelines were used to develop the tabular format shown in Figure 4-1.

- Three spaces below the text, *table* is written in capital letters and is

TABLE I

A CLASSIFICATION OF MATERIALS USED TO DRILL ROCK
CLASSIFIED IN DESCENDING ORDER OF HARDNESS
ACCORDING TO THE KNOOP SCALE

| Materials | Kilograms per square millimeter |
|---|---|
| Diamonds | 7000 |
| Boron carbide | 2750 |
| Aluminum oxide | 2100 |
| Tungsten carbide | 1880 |
| Steel | 800 |

Source: Harold Bridwell, research engineer.

Table I shows that diamonds are almost nine times harder than steel.

FIGURE 4-1  An example of formal classification

followed by a roman numeral. The table number is centered two spaces above the classification title.

- A complete title that contains at least four parts introduces the classification: (1) the general class of items being classified, (2) the limitation of the class, (3) the method used for arranging items, and (4) the basis used for the classification. (The basis in this table is hardness.) Any additional information needed to explain the classification must also be included in the title. In Figure 4-1, "according to the Knoop Scale" was added because hardness is determined in more than one way and various scales are used.

- All letters in the title are capitalized. The title, which is single-spaced, does not extend beyond the right-hand or left-hand edges of the classification. If the title requires more than one line, and it usually does, the lines should vary in length; and each line is centered above the classification. For suggested line arrangements, see Figure 7-2.

- Two spaces below the title, a solid horizontal line is drawn. Two or three spaces below the first line and parallel to it, a second line is drawn. (Enough space is left between the two lines to permit subtitles to be written.)

- One space below the last entry in the classification, a third horizontal line is drawn.

- Two spaces below the bottom line, a source is given if the report writer has used another person's classification or if the writer wants to receive recognition for the work. When a table has been originated by the writer, the source may be designated as *primary*.

- Summarizing statements are always used with a table or other graphic aid (Unit 5). As Figure 4-1 shows, this text begins three spaces below the completed table.

The classification in Figure 4-1 can be presented as text only.

**EXAMPLE**

The Knoop scale for hardness of material used to drill rock is based on kilograms (kg) per square millimeter (sq mm). Five materials classified in descending order of hardness according to this scale are diamonds, 7000 kg per sq mm; boron carbide, 2750 kg per sq mm; aluminum oxide, 2100 kg per sq mm; tungsten carbide, 1800 kg per sq mm; and steel, 800 kg per sq mm.

The same classification in Figure 4-1 can also be presented as text combined

The Knoop Scale for hardness of material used to drill rock is based on kilograms (kg) per square millimeter.  Five materials classified in descending order of hardness according to the Knoop Scale are shown in the bar graph, Figure 3.

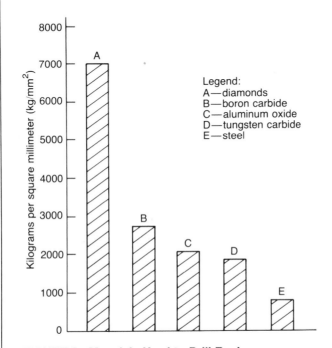

FIGURE 3.  Materials Used to Drill Rock

Figure 3 shows that diamonds are more than seven times harder than steel. They are also more than twice as hard as any other rock-drilling material.

**FIGURE 4-2**  An example of a formal classification using text combined with an illustration

with a bar graph, as shown in Figure 4-2. A summarizing statement similar to the one given for Table I is written three spaces below the figure title.

Occasionally, a formal classification, like a limited classification, contains subdivisions. The items being classified are then arranged in outline form, as shown in Figure 4-3. As indicated in Figure 4-3, the basis relates only to the

### TABLE II
#### THE CENOZOIC GEOLOGICAL ERA CLASSIFIED
#### ACCORDING TO THE TIME IN YEARS SINCE
#### THE BEGINNING OF EACH EPOCH

| Cenozoic Era | Years since beginning of each epoch |
|---|---|
| I. Quaternary Period | |
|    A.  Recent Epoch | 50,000 |
|    B.  Pleistocene Epoch | 1,000,000 |
| II. Tertiary Period | |
|    A.  Pliocene Epoch | 12,000,000 |
|    B.  Miocene Epoch | 30,000,000 |
|    C.  Oligocene Epoch | 40,000,000 |
|    D.  Eocene Epoch | 60,000,000 |

Source:  Primary.

**FIGURE 4-3** An example of a formal-outline classification

items in the subclass. Again, including more than one subclass is not recommended.

## SUMMARY

1. Classification, a systematic arrangement of items that have one or more common traits, is a valuable tool for technical writers. It can organize, graphically portray, and simplify text in technical reports.

2. Classification used in technical communication is divided into two general kinds: *unlimited* and *limited*.

3. Unlimited classifications are merely long or short lists of related items organized systematically under a meaningful heading. Only short unlimited classifications add clarity to a report. An appropriate heading can usually keep the classification short.

4. Limited classifications are divided into two kinds: informal and formal. When

headings, lead-in sentences, or topic sentences limit them to pertinent information, both kinds contribute significantly to communication.

5. A formal classification can be presented as a table and text, as text, or as a figure and text. *Note:* All graphic aids (Unit 5) require text to introduce and summarize them.

6. Writers are obligated to limit all classifications they use to include only data needed for a specific purpose.

7. Classifications are particularly important in analytical reports (Unit 10). They can also help clarify data in various other kinds of reports such as descriptive, process, and proposal.

8. Partition, the division of a single item into its parts, is used in most physical descriptions (Unit 8).

UNIT

## SUGGESTED ACTIVITIES

A. Using formal or expanded definition, define the following terms as you understand them. If necessary, include examples, illustrations, or other techniques discussed in Unit 3.

   1. Partition
   2. Classification
   3. Unlimited classification
   4. Limited classification
   5. Formal classification

B. In a table, use the following information to write a formal classification of nickel alloys, showing the percentage of nickel contained in each alloy: monel, 67 percent nickel; "k" monel, 66 percent nickel; "z" nickel, 94 percent nickel; inconel, 80 percent nickel; hastelloy A, 56 percent nickel. Compose an appropriate title for the classification. Above the table, write a sentence that introduces the table to the reader. Below the table, write one or more statements that draw conclusions from the table.

C. Prepare an outline or a table for either a limited or a formal classification for a subject that can be classified. Select a subject related to technology and one that is familiar to you. Write an introduction for the classification, present the classification, and discuss the important concepts illustrated.

D. Find a classification in a textbook or other printed material related to technology. Write an evaluation of the classification based upon the following questions:

   1. Is the title complete? Does it state specifically what the reader should learn from the classification?
   2. Is the classification simple enough to be easily read and interpreted?
   3. Is the information systematically arranged?
   4. Is the classification coordinated with the text?
   5. Is the classification useful or is it a distraction?
   6. Are symbols or abbreviations used in the classification? If so, are they meaningful to you as the reader?
   7. Is the classification appealing to the eye?
   8. Is the information factual and specific enough that it can be interpreted in only one way?

# Graphic Aids

## OBJECTIVES

After reading this unit, you should be able to do the following tasks:

- List and describe the uses for graphic aids in reports.
- List and describe the kinds of graphic aids that can be used.
- Integrate graphic aids into the text of a report.

## OVERVIEW

In technical writing, a graphic aid is any table, diagram, or picture used in a report to clarify written information. Some writers design graphic aids manually; others use computers. A detailed study of graphic aids is complex. Technicians who want to learn the specific uses for specific aids and the procedure for developing graphic aids can find this information in technical illustration courses. This unit is limited to three topics that explain the relationship between graphic aids and written information in technical reports: (1) the uses for graphic aids in reports, (2) the four general classes of graphic aids, and (3) the correct procedure for integrating graphic aids with the text.

# USES FOR GRAPHIC AIDS IN REPORTS

A graphic aid is never an isolated part of a technical report. It is used to supplement the text in one or more of three specific ways:

1. To organize quantitative information that cannot be clearly explained in words
2. To clarify unfamiliar technical terms and processes
3. To emphasize important details concerning trends, characteristics, concepts, and relationships of one thing to another.

## Organizing Quantitative Information

Quantitative information is any detailed subject matter that concerns statistics, percentages, rates, characteristics, or similar data. A written paragraph concerning quantitative information can be boring, confusing, and easily forgotten because specific facts are interwoven with often complicated verbal explanation.

EXAMPLE

The melting point of olive oil is 2 degrees Fahrenheit; the boiling point is 464 degrees Fahrenheit. The melting point of glycerin is 64 degrees Fahrenheit; the boiling point is 554 degrees Fahrenheit. The melting point of castor oil is 10 degrees Fahrenheit; the boiling point is 595 degrees Fahrenheit.

Even the small amount of specific information in this paragraph is buried among words. It can be emphasized and quickly interpreted when introduced as a table. See Figure 5-1.

In the table in Figure 5-1, because the numbers are separated from the words, both the numbers and the words are easily understood. A detailed explanation of tables is discussed under "Kinds of Graphic Aids" later in this unit.

## Clarifying Technical Terms and Processes

As explained in "Expanded Definition," Unit 3, a reader can more clearly understand a description of a technical term or process if the description includes a graphic aid that has been carefully selected to supplement the written explanation. See Figure 5-4 presented later in this unit.

## Emphasizing Important Information

A diagram, picture, or graph can increase the emphasis a writer wants to give significant trends, characteristics, or concepts related to a report topic. Again,

**TABLE I**

**MELTING AND BOILING POINTS**

**OF THREE LIQUIDS**

**EXPRESSED IN DEGREES FAHRENHEIT**

| Liquid | Melting point | Boiling point |
|---|---|---|
| Olive oil | 2 | 464 |
| Castor oil | 10 | 595 |
| Glycerin | 64 | 554 |

**FIGURE 5-1**  A sample table

the writer needs to apply judgment (Unit 2) to determine whether a particular graphic aid contributes to or minimizes the written text.

# KINDS OF GRAPHIC AIDS

Classifying graphic aids can be a complicated procedure. In this unit, graphic aids are divided into four general types:

1. Tables
2. Diagrams
3. Pictures
4. Graphs

The definition, use, and format for each type are explained.

## Tables

Tables include any graphic aids in which information is arranged in rows and columns. Tables are used primarily for quantitative information that must be precise. The information usually involves amount of money, percentages, weights, measurements, and similar uses of numbers. Tabular information may also list characteristics of materials or analyses. For example, a table of cast-iron welding procedures lists the types of cast iron used in welding and then tabulates the properties of and the procedure for treating and welding each type. Similarly, the properties and uses for rocks, minerals, and other materials may be organized into a table. A table usually has two or three columns of tabulated information related to the listed topics. If it is used for formal classification, however, it has only one column of tabulated information.

All tables are arranged in vertical columns usually separated by vertical lines. The entire table may be enclosed in a box, or the end lines of the box may be omitted. The table number and the title are placed above the box, as illustrated in Figure 5-2.

### TABLE II

#### DISTRIBUTION OF EARNINGS OF FULL-TIME EMPLOYEES
#### IN THE UNITED STATES IN MARCH 1984
#### SHOWN SEPARATELY IN THOUSANDS FOR MEN AND WOMEN

| Annual earnings | Number of men in thousands | Number of women in thousands |
|---|---|---|
| Less than $3,000 | 954 | 555 |
| $3,000 to $4,999 | 446 | 429 |
| $5,000 to $6,999 | 918 | 1,305 |
| $7,000 to $9,999 | 2,480 | 3,747 |
| $10,000 to $14,999 | 6,076 | 7,932 |
| $15,000 to $19,999 | 6,949 | 5,597 |
| $20,000 to $24,999 | 6,644 | 2,891 |
| $25,000 to $49,999 | 14,206 | 2,547 |
| $50,000 and over | 2,856 | 165 |
| Total | 41,592 | 25,168 |

Source: Department of Commerce, Bureau of the Census.

FIGURE 5-2  An example of table format for technical reports

The following guidelines relating to Figure 5-2 are given to help the report writer understand the table format used in technical reports.

- The table number is centered above the box. All letters in the word *table* are capitalized. A roman numeral follows the word *table*.

- Two spaces below the table number, the title of the table is centered and written entirely in capital letters. The title is centered above the table and is indented at least five spaces from the left-hand and the right-hand edges of the table. If two or more lines are needed for the title, each line is single-spaced and centered. For suggested line arrangements for titles, see Figure 7–2. The horizontal line shown below the title is optional; however, two spaces always separate the title from the table. A column title is written above each column. A horizontal line and two spaces separate this title from the column.

- The topics are listed in the left-hand column. Quantitative information is placed in the other columns of the table. Tables should be simple. More than three columns of quantitative information can be confusing, especially when the list of items is long. Nevertheless, tables of several columns are frequently seen in reports currently being written. When complex tables are being considered, a writer should decide whether all the information is needed; whether the table is more distracting than informative; whether all readers, as opposed to most or some or few, will benefit from studying the table; and whether the table can be effectively divided into two or more tables so that the presentation of information can be simplified.

- If the information in the table must be totaled, a horizontal line, usually double, separates the totals from the entries and extends across the entire table. Two spaces below this line, the word *total* is placed in the left-hand column and the first letter is capitalized. The total for each column is listed opposite the word *total* in the appropriate column.

- Two spaces below the box on the left-hand margin of the table, the word *source* is written to identify the originator of the table and is followed by a colon. The first letter of *source* is capitalized. As previously stated, if the table is the writer's own work, *source* may be followed by the word *primary*, or the source note may be omitted.

## Diagrams

Diagrams are usually classified into four types: block diagrams, pictorial diagrams, schematic diagrams, and wiring diagrams. All diagrams, however, are used for one specific purpose in technical reports: to assist the writer in explaining how the various parts of a complex mechanism help make it work.

In technical reports, one format is used for all four kinds of diagrams. This format is illustrated in the diagram of a transformer circuit, Figure 5–3.

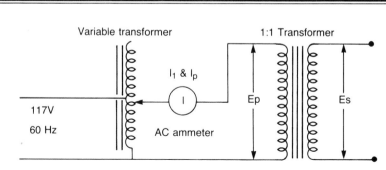

**FIGURE 1. Transformer Coupling Using a Variable Transformer to Vary the Voltage to a 1:1 Transformer Under Experimental Analysis**

**Source:  Richard W. Tinnell, <u>Electronics Electricity</u> (New York: Delmar Publishers, 1971), p. 161.**

**FIGURE 5-3** An example of format for diagrams

The following guidelines explain the format used in the illustration.

- Diagrams are never enclosed boxes.

- Diagrams are referred to as figures. The word *figure* is written as a complete word two spaces below the lowest part of the diagram. The first letter of *figure* is capitalized and is placed on the left-hand margin of the diagram. The word *figure* is followed by an arabic numeral and a period. The numbers for diagrams are written consecutively with numbers for all other graphic aids used in the report except tables. For example, if a graph and a photograph precede the diagram, the diagram is Figure 3.

- A title follows the figure number. The first letter of all important words in the title is capitalized. If the title extends beyond the right-hand margin of the diagram, the second line of the title begins one space below the first line at the left-hand margin.

- Symbols and standard abbreviations are used to identify most parts of a complex schematic diagram; however, a large component such as an oscillator, transformer, or motor may be identified by its name. Complete words are usually written within the rectangular boxes of a block diagram. A firm rule for the use of symbols, abbreviations, and words on diagrams cannot be made. Only the person originating the diagram can determine the methods of labeling parts that will result in clarity and neatness and, above all, will not confuse the reader.

- The source of the diagram is written two spaces below the last line of the title, as illustrated in Figure 5-3. Also see Figure 5-2.

## Pictures

Pictures include photographs and drawings. They are used in technical reports to clarify or emphasize verbal explanations or descriptions. An exception is the use of cutaway drawings, drawings in which part of the outer cover of a device or mechanism is removed to expose inner parts. They may be classed as diagrams. The format for pictures in technical reports is identical to the diagram format discussed and illustrated in Figure 5-3.

## Graphs

Graphs, sometimes referred to as charts, are an arrangement of lines, bars, symbols, or shading that represent successive changes in a variable. In technical writing, graphs are used primarily to indicate trends or comparisons. Therefore, a graph is sometimes substituted for a table when showing how certain facts interrelate is more important than tabulating facts. In contrast to some tables, a carefully prepared graph has only two variables: an independent variable written horizontally, and a dependent variable written vertically; see Figure 5-4. An *independent variable* is a variable chosen by the person who is gathering information. A *dependent variable* is the variable over which the person who is gathering information has no control.

The following guidelines relating to Figure 5-4 explain the format used for all graphs in technical reports.

- The dependent variable is always vertical. The independent variable is always horizontal.

- All graphs except pie graphs are rectangular. Sometimes, a rectangle surrounds them. Normally, however, only a left-hand vertical line and a horizontal baseline (Figure 5-4), used for data, are drawn. Outside the left-hand margin, dependent variables are arranged vertically. Below the baseline, independent variables are arranged horizontally. The format for pie graphs is identical to the one used for diagrams.

- A legend (a list that explains all symbols and abbreviations) may be included. Symbols and abbreviations, however, are used only when limited space prevents the use of words. A required legend is placed in any

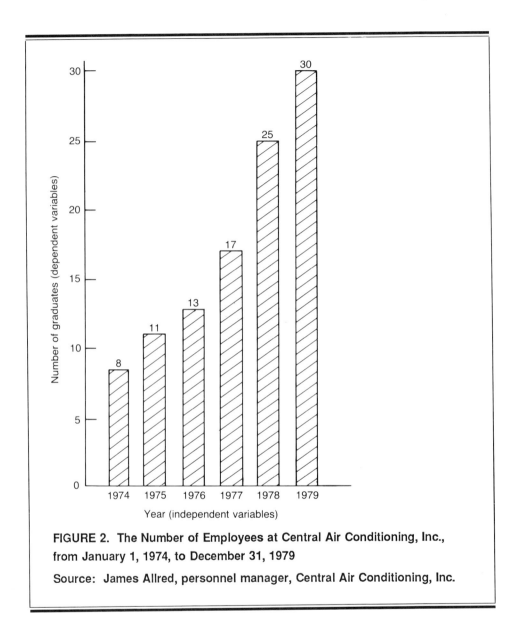

FIGURE 5-4 An example of format for graphs

space within the rectangular format that results in an attractive graph. The graph used as an illustration in Figure 4-2 shows one legend format.

■ The figure number, title, and source format for graphs is identical to the figure number, title, and source format used for diagrams and pictures, as illustrated in Figure 5-4.

# INTEGRATION OF GRAPHIC AIDS WITH THE TEXT

Graphic aids are used in technical reports for only one reason: to supplement the text so that the reader has a complete understanding of a detailed subject. Graphic aids, therefore, must appeal to the reader so that they are understood and used. They must also be coordinated with the text so that the subject evolves systematically because of them.

## Appealing to the Reader

The following guidelines are offered to help a writer design graphic aids that appeal to the reader.

- The reader's interests and experience must be considered. Does the reader like graphic aids? Can he or she interpret them easily? What kinds of graphic aids are most meaningful to the reader?

- Both the top and the bottom of the graphic aid are separated from the text by three spaces. The graphic aid includes everything related to it — table or figure number, title, and source.

- The graphic aid must not be crowded into a limited amount of space. Space is determined by the kind of graphic aid being used and the amount of information it contains. White space around each item within the graphic aid gives an attractive, organized appearance that increases clarity.

- The graphic aid must be neat and distinct. Lines must be straight, even, unsmeared, and dark enough to be seen easily. Words and abbreviations must be neatly printed or typed and placed near the parts they identify. Letters within the words must be evenly spaced.

- A decision concerning the use of symbols, abbreviations, or complete words depends on the reader's understanding of them, the amount of space available within the graphics aid for identification, and the number of identical parts such as resistors and capacitors that are used in the graphic aid. As mentioned earlier, a legend is included for any symbols or abbreviations that may be unfamiliar to a reader.

- A graphic aid contains only facts that have been researched and can be verified.

- A complex table, graph, or diagram is difficult to interpret. Two or more simply constructed graphic aids are often preferable to a complex one.

## Coordinating Graphic Aids with the Text

The following guidelines help a writer coordinate a graphic aid with the text so that the graphic aid and text supplement each other.

- Each graphic aid must have a number. The number is used each time the graphic aid is discussed in the text. Referring to a table or figure as "the foregoing table" or "the following chart" is not specific. A table or figure that is not on the same page as the reference to it is identified by both the table or figure number and the page.

**EXAMPLE**

**The drawing in Figure 2, page 9, shows a bearing-plate assembly.**

- When a particular figure is mentioned or discussed in technical reports, the figure is identified by its specific name such as block diagram, drawing, photograph, or bar graph. (See the foregoing example.)

- The title of a table or figure must be complete enough to tell the reader the kind and amount of information given in the graphic aid. (See Figures 5–1 to 5–4 for examples of complete titles.)

- Unless the writer is certain that a graphic aid contributes to the purpose of the report, it should be omitted. Unnecessary graphic aids distract and confuse the reader.

- Preferably, a graphic aid immediately follows an introductory statement that refers the reader to the illustration. If space at this point is not available, the graphic aid is presented as soon as possible after it has been introduced.

- A graphic aid is both introduced and discussed. The writer is obligated to explain the relationship between the graphic aid and the text. An effective method of introducing a graphic aid and discussing the content is given in Figure 5–5.

**The flow chart in Figure 1 shows how copper taken from a mine is prepared for marketing.**

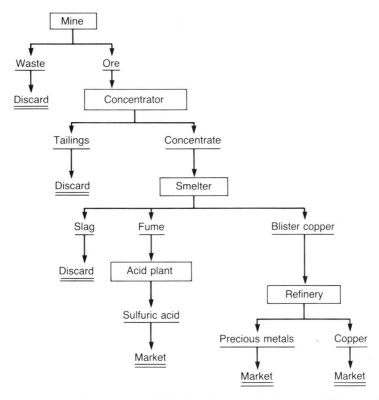

**FIGURE 1.  The Process of Moving Copper Ore from the Mine to the Market**

**The flow chart shows that, at the mine, waste is separated from the copper ore and discarded.  The ore is processed through a concentrator, a smelter, and a refinery.  In each process, additional impurities are separated from the ore until copper that is 99.99 percent pure reaches the market.  Two by-products, sulfuric acid and precious metals, are also marketed.**

**FIGURE 5-5** An example of coordinating a graphic aid with the text

- The value of placing graphic aids in an appendix to a report is questionable. Important graphic aids belong in the text. Unimportant ones should be omitted.

▪ A graphic aid should be limited to one page. However, if more than one page is required, a foldout sheet is used. A graphic aid is never divided into two or more parts.

# SUMMARY

1. A graphic aid is any kind of table, diagram, or picture used in technical communications to clarify text.

2. Graphic aids have three primary functions: first, to organize information into a form that is more meaningful than words; second, to add clarity and emphasis to written text; and third, to visualize such information as trends, characteristics, and relationships of one thing to another.

3. In general, tables most effectively present quantitative data. Charts, diagrams, pictures, and graphs most effectively show trends, comparisons, and contrasts.

4. The topics "Appealing to the Reader" and "Coordinating Graphic Aids with the Text," pages 53 and 54, offer several important guidelines to help a writer prepare easily interpreted, meaningful, and interesting graphic aids.

5. One guideline related to graphic aids is basic: Graphic aids are an extension of text, used to support the written message. Therefore, a writer always introduces them before presenting them. He or she also uses a summarizing statement, preferably following the aid, to point out the main idea being illustrated.

6. A graphic aid has little value unless it appeals to the readers, the readers understand it, the information is complete, and the format is attractive.

7. Two or more simple graphic aids are usually better than a single complex one. As in other procedures related to technical writing, judgment plays an important role in the choice and preparation of graphic aids.

UNIT

# SUGGESTED ACTIVITIES

A. In a textbook or in supplementary material related to technology or to your main field of study, find a graphic aid and use it to answer the following questions:

1. Was the graphic aid introduced to the reader? If the answer is yes, tell how. If the answer is no, write an introduction.
2. Did the writer coordinate the graphic aid with the text? If the answer is yes, tell how. If the answer is no, tell how it could have been coordinated.
3. Is the information in the graphic aid pertinent and complete? Justify your answer.
4. Is the information easily interpreted? Give specific reasons to justify your answer.
5. Does the graphic aid appeal to you as a reader? Give specific reasons for your answer.
6. Is the title complete and meaningful? If it can be improved, explain how.
7. How are the parts of the graphic aid identified—by symbols, abbreviations, complete words, or a combination of two or more of these? Are the identifications easy to understand?
8. Is the graphic aid necessary? Justify your answer.

B. Prepare a table of quantitative information concerning a subject familiar to you. Identify the table as Table I, prepare an informative title, introduce the table, present the table, and write a summary of important conclusions derived from the table.

C. Select a technical subject that requires the use of a graphic aid other than a table to give the reader a complete understanding of the subject. Use a subject familiar enough to you that you can make it meaningful to the reader. Introduce the graphic aid to the reader, present the graphic aid, and then discuss it. In the discussion, avoid a detailed explanation. The discussion should be a general one related to the detailed information in the graphic aid. Use the rules given in this unit to create an original one. Do not use or duplicate a graphic aid designed by someone else.

# The Outline

After reading this unit, you should be able to do the following tasks:

- List the advantages of an outline.
- List and describe the three divisions of an outline.
- Prepare an outline for a report.

## OVERVIEW

A technician who constructs a piece of equipment must follow a diagram, schematic, or print. Such aids show the organization and interaction of parts that permit the completed equipment to function efficiently. An outline functions as the writer's diagram or schematic. It organizes parts of a report so that the central theme is developed logically and clearly.

Before a writer begins to develop an outline, he or she shows what the report related to it is supposed to accomplish — in other words, the purpose of the report. He or she also has collected the information to be included and has established a possible plan for developing this information. Once the writer has clearly identified the purpose and a possible development plan, preparing an outline can save him or her time, money, and energy.

This unit explains the advantages of an outline, describes the components of an outline, and gives suggestions for developing an outline.

# ADVANTAGES OF AN OUTLINE

Quality writing results from exact and orderly thinking. An outline is a valuable tool that helps writers get their thoughts organized. Here are some advantages of making an outline.

1. An outline, like a schematic, is a systematic arrangement of parts. In planning an outline, a writer learns to organize thoughts.
2. The outline organizes report topics in a logical sequence. The finished report is a unit of information, not a list of isolated facts.
3. A writer who develops a complete outline gathers all the facts necessary to accomplish a predetermined purpose. Thus a detailed outline helps ensure that pertinent information will not be omitted from the report.
4. An outline planned to accomplish a single purpose helps the writer recognize the nonessential information that should be omitted from a specific report.
5. An outline encourages a writer to develop a straightforward writing style. This style tends to reduce errors in punctuation, grammar, and sentence structure.
6. An outline is an excellent guide for the table of contents frequently used in long, formal reports.
7. An outline indicates what topics are to be used as center headings, margin headings, and paragraph headings in the report.
8. An outline, because it shows the logical changes in topic discussion, helps the writer make smooth transitions from one part of a report to another.
9. An outline eliminates the need for extensive revision of a report. Because of the logical development, most changes in the first draft of the text are word and phrase changes rather than major changes in thought, organization, or subject matter.

# THE DEVELOPMENT OF AN OUTLINE

One basic outline, with slight modifications, can be used for every kind of technical report. Every report outline has three main divisions:

1. An introduction
2. A body
3. A summary

Each division evolves logically and meaningfully from the preceding one. The introduction defines the complete report. The information defined in the introduction is detailed in the body of the report. The summary restates what has been said in the body and, when necessary, states conclusions and recommendations. See the basic outline illustrated in Figure 6–1.

---

### A BASIC OUTLINE FOR REPORTS

I. Introduction
   A. Background (always used)
   B. Purpose of the report (always used)
   C. Source (or sources) of data (used only when needed)
   D. Justification for omitted data (used only when needed)
   E. Definitions (used only when needed)
   F. Major topics (always used)
      1. First major topic discussed in the body
      2. Second major topic discussed in the body
      3. Third major topic discussed in the body
II. Major topic No. 1 (taken from the introduction and developed in detail)
   A. Subtopic to major topic No. 1
      1. Details of subtopic A
      2. More details of subtopic A
   B. Second subtopic to major topic No. 1
      1. Details of subtopic B
      2. More details of subtopic B
         a. Details of 2
         b. More details of 2
III. Major topic No. 2 (taken from the introduction and developed in detail as illustrated in major topic No. 1)
IV. Major topic No. 3 (taken from the introduction and developed in detail as illustrated in major topic No. 1)
V. Summary
   A. Recapitulation (always used)
   B. Conclusions (used only when needed)
   C. Recommendations (used only when needed)

---

**FIGURE 6-1** An example of a basic outline for technical reports

## Developing the Introduction

As illustrated in Figure 6-1, the introduction consists of three to six parts (A–F). The following discussion explains each part.

1. Part A, the background, is the first item in all outlines and reports. It is often made up of only one or two sentences and rarely exceeds a short paragraph. It explains why a report is being written and also prepares the reader for the statement of purpose that immediately follows. Both the background and the purpose of an analytical report are illustrated in the following example.

**EXAMPLE**

**For three years, Union Press has been considering the advantages of converting from hot type to cold type for printing its publications. This report analyzes both kinds of type to determine the advantages of each and the cost of conversion.**

The first sentence, the background, explains why the report is being written. The second sentence explains the purpose, the goal to be attained in the report.

2. Part B, the purpose, is an essential topic in every outline and report introduction. Always stated specifically, clearly, and completely, the purpose determines the following: what information is to be included in the report, how the information is to be arranged, and how much detail is required. Thus, every word, sentence, and paragraph in a report is aimed toward accomplishing the goal expressed in the purpose.

3. Part C, the source (or sources) of data, is needed any time facts or figures used in the report must be justified. For example, a writer who states that a piece of equipment costs $2750 must prove that statement.

4. Part D, the justification for incomplete data, is used when a writer must submit a report, even though all pertinent data is not available. For example, one of three required estimates has not been received before a report submission deadline; the writer must justify the omission.

5. Part E, the definition section is used when a short list of formal definitions or a single expanded definition is needed. The importance of and guidelines for using definitions have been explained in Unit 3.

6. Part F, a listing of the major topics, is included in every outline introduction. One basic rule for technical writing states that no subject can be discussed until it has been introduced. Logically, the major topics pertaining to the report subject are listed or briefly explained immediately before the detailed text is presented.

## Developing the Body of the Outline

After the introduction to a report has been developed, most of the outline is complete. The purpose has been determined; explanations concerning data, if needed, have been made; required definitions have been given; and major topics have been logically arranged. The second division of the outline, the body, is used to develop each major topic in enough detail to accomplish the purpose of the report defined in Part B of the introduction.

The term *body* is excluded from the outline. In its place, each major topic is listed as a main heading and is preceded by a Roman numeral beginning with II. See Figure 6-1.

In typical outline arrangement, as shown in Roman numeral II of the basic outline, detailed information concerning each topic is subdivided until all facts needed for the report have been included in the outline. This detailed subdivision of information helps the writer differentiate between general concepts and specific information. It also helps the writer develop each topic systematically and thus avoid needless repetition and wordy generalizations.

## Developing the Summary

Summary, the third division of an outline, always begins with a recapitulation of the ideas emphasized in the report. This recapitulation is brief; it never includes new information. In a few sentences or a paragraph, it shows the reader how the purpose of the report has been accomplished. Each of the sample reports in Units 8 through 12 has a recapitulation.

Conclusions (sometimes called interpretations) and recommendations often follow the recapitulation, as shown in the analytical report in Unit 10. The conclusions are the inferences made by the writer from the information presented in a report. The recommendations, based upon the conclusions, are the writer's suggestions for some kind of action.

# PREPARATION OF AN OUTLINE

An outline, which is primarily a tool for the report writer, is rarely attached to a completed report. The following guidelines help a writer develop a useful outline.

- A topic outline is one in which each main topic and each subdivision topic is identified by words or phrases. For example, Figure 6-1 shows a topic outline. Some writers prefer a sentence outline, one in which each topic is written as a complete sentence. Often, a writer uses topics and supplements them with sentences; the topics are listed and then followed by a lead-in sentence in parentheses after the words or phrases. A writer who understands the purpose of outlining may ultimately develop a useful technique different from the one discussed above. In fact, a few writers find that a rough draft of the report, rearranged with scissors and tape after it is completed, is their best approach. A good organization plan for any writer is the one that helps him or her write a clear, concise, complete, logically developed report in the least amount of time.

- A person who prepares a complete outline may save at least half the time and effort required to write a report from a sketchy list of major topics. This saving benefits both the employer and the employee.

- Using parallel structure for items that are equal to one another in importance requires a little time and thought, but parallel structure in the outline aids in logical development of information in the report. The following example illustrates parallel structure used for the major topics necessary to explain the process of preparing a contour map.

**EXAMPLE**

I. **Taking the readings**

II. **Interpreting the results**

III. **Plotting the data**

Parallel structure is also used for all subtopics listed under a major topic.

- The words or phrases used to identify topics in an outline should be meaningful enough to be used later as informative headings in the report. For example, saying "Preparing the mix" is better than saying "Preparation" or "Mix."

- A neat outline that clearly indicates major topics and the subdivisions of the major topics simplifies heading arrangements in a report.

- Checking an outline a few hours after it has been prepared helps a writer determine whether all information in the outline is pertinent, whether it is complete, and whether it is logically arranged.

# SUMMARY

1. An outline is a plan that helps a writer organize the various parts of a report into a logical sequence of information.

2. A writer can develop a useful outline only after he or she knows the purpose to be accomplished in the report, has accumulated all the information needed for the report, and has a tentative mental plan for development.

3. Developing an outline offers several advantages to a writer who prepares one before writing a report.

4. All effective reports have an introduction, a section for presenting detailed information, and a summarizing section. The topic outline illustrated and discussed in this unit shows the kind of data logically contained in each section.

5. The basic outline in Figure 6–1 illustrates a plan that has proved to be helpful to beginning writers. Through trial and error, a writer may later develop the plan he or she prefers.

UNIT

## SUGGESTED ACTIVITIES

A. Identify the part of an introduction to an outline that is illustrated by each of the following statements:

1. Because ABC Company is expanding its operations, larger facilities must be acquired.

2. Making steel in a Bessemer converter requires four major operations: (1) putting the steel into the converter, (2) heating the steel, (3) removing the slag from the converter, and (4) removing the steel from the converter.

3. The estimated costs for constructing the new building were obtained from three local contractors: Mrs. James F. Thomas, Mrs. Elaine B. Franks, and Mr. Ward D. Friedman.

4. Hot-pressing is a process in which heat and pressure are applied simultaneously to a material to produce a polished surface.

5. This report explains the operation of a worm-and-wheel assembly.

6. A bevel gear is a gear wheel that is meshed with a second gear wheel so that their shafts are at an angle of less than 180 degrees.

7. Three locations for a new plant are analyzed in this report. The analysis is based upon the construction costs, maintenance costs, and access to railroad and trucking facilities.

8. The data concerning the cost of materials is incomplete because cost figures have not been received from the Kirch Lumber Company.

9. A direct-current motor has three basic parts: (1) field poles, (2) an armature, and (3) a commutator.

B. Find a technical report in a textbook, a technical library, or a technical magazine. Sometimes, a company will permit a student to duplicate a report written by one of its employees. Determine whether the report was developed according to a logical outline. Make notes concerning the data contained in the introduction, the body, and the summary. Then answer the following questions and be able to explain your answers with specific references:

1. Can the organization be improved?
2. Has all essential data been given?
3. Has any data been needlessly included or repeated?
4. Can you easily understand the text?

# Format

After reading this unit, you should be able to do the following tasks:

- Discuss and use the guidelines for producing an effective text format.
- Describe the three divisions of a report title.
- Discuss and use the guidelines for planning and writing effective report headings.

## OVERVIEW

A reader may form a favorable or an unfavorable attitude toward a report by its general appearance. Therefore, the importance of the format cannot be overemphasized.

In Unit 1, where correct format was introduced as a basic principle of technical writing, format was defined as the arrangement of headings, text, and supplements in a report. The format for three supplements— definition, classification, and graphic aids — has already been presented. This unit is concerned with the format for the text, the report title, and the topic headings. The format for introductory supplements to the report is explained in Unit 16.

# TEXT FORMAT

Text format refers to the general makeup of a typewritten or printed page within a report. A writer who has access to both a word processor and the recently marketed electronic publishing system can readily create any desired format. He or she can use any margin or line arrangement to type all information onto the word processor. The publishing system then can manipulate both text and graphics until the desired format is created.

One disadvantage in using this versatile electronic system is that writers may be inclined to make a format so spectacular that it detracts from the basic reason for writing the report— to objectively transmit information to a reader. If each page of the text is arranged in accordance with the eight following guidelines, the completed report will be neat, attractive, and easy to read.

- All formal reports are typewritten on white, high-quality typing paper that measures 8 1/2 inches by 11 inches, is 15- to 20-pound bond, and has a minimum fiber content of 25 percent. The text has no typographical or spelling errors. Erasures are made so that the erased error is not obvious; however, word processors and the recently developed electronic typewriters are eliminating the need to erase.

- All report text is double-spaced, and each paragraph is indented five spaces from the left-hand text margin. Quoted material extending beyond three lines is the one exception to this rule. It is single-spaced, separated from other text by a double space, and indented five spaces from both the left-hand and the right-hand margins. The beginning of a paragraph is indented an additional five spaces. Quotation marks are not used. The following example shows this kind of quotation.

**EXAMPLE**

In *Elements of Data Processing,* page 68, Ellen Marxer explains a magnetic disk:

The magnetic disk is a thin, closely machined aluminum disk. Some manufacturers coat disks with a magnetic oxide coating, and others plate the disks with a nickel-cobalt compound. The manufacturer who plates with nickel-cobalt achieves a greater bit-and-track density on the disks.

Data is stored as magnetic spots on tracks on each surface of the disk, much like the magnetic spots on magnetic tape.

Note that paragraphs are single-spaced unless the quoted text is taken from more than one source. Quoted text three lines or less in length is enclosed in quotation marks and included in the general text.

- The first page of text begins 2 inches from the top of the paper; other pages begin 1 inch from the top. All pages should have a 1 1/2-inch left-hand margin, a 1-inch right-hand margin, and a 1-inch bottom margin. The extra 1/2 inch on the left allows for binding the report or enclosing it in a folder. A backing sheet placed behind the page to be typed simplifies the process of typing an attractive report; See Figure 7–1.

  Figure 7–1 shows the margins to be used for the text, the 2-inch margin at the top for the first page of text, the location for the page number in the upper right-hand corner, the 1-inch margin at the top for the second page and subsequent pages, and the number of lines of text that may be written double-spaced on one page. The numbers on the lower left-hand margin indicate the number of single lines available for footnotes, if used. See Unit 17 on page 196. The center horizontal and vertical lines, used for arranging a title page, are explained later.

  Backing sheets can be used only for typed text. Writers using word processors must make margin and line adjustments needed to comply with the foregoing guidelines.

- Arabic numeral 1 is centered below the text on the *first page* of the report. All other pages are numbered consecutively, beginning with Arabic numeral 2, above the text in the upper right-hand corner.

- Neither classifications, graphic aids, nor written information extends beyond the designated left-hand or right-hand margins. The bottom margin may vary occasionally. See the last two guidelines.

- A writer should attempt to write a complete word at the end of a typewritten line. When the word must be divided, a hyphen is used according to the following rules:

  1. Words are divided only between syllables. All standard dictionaries show the syllables of words.
  2. Long, one-syllable words like *through*, *strength*, or *ground* cannot be divided. This rule also applies to *dropped*, *stopped*, *formed*, *climbed*, and similar past-tense forms of verbs in which the suffix does not create a second syllable.
  3. A word containing fewer than seven letters should not be divided.
  4. The use of fewer than three letters at the end of a line is not recommended.
  5. A number stated in figures cannot be divided. For example, the number 4,572,625 must be written on one line.

Page number  0

1

2

Top margin for title | and first page

3

4

5

6

7

8

9

10

11

12

13

14

15

16

17

18

19

20

21

12 —————————————————    22
11
10                                          23
9
8                                           24
7
6
5                                           25
4
3                                           26
2
1                                           27

**FIGURE 7-1**  An example of a backing sheet for report text and title page

- The first line of a paragraph is not placed on the last line of a page.

- An incomplete line of text or the last line of a paragraph is not placed at the top of a page.

# THE REPORT TITLE

Report titles vary from one another in content, length, and format. The content of a report title determines how the title is arranged. The length of the report determines whether the title is placed on the first page of the text or on a title page that precedes the text.

## Planning a Report Title

A report title is written not to arouse a reader's curiosity but to introduce the purpose of the report. Planning an effective, informative title is simplified if the following guidelines are observed:

- A title always specifies, first, the type of report being written (a physical description, a process, an analysis, or some other type); second, the subject of the report; third, the limits of the information.

**EXAMPLE**

**The Process Used at Sunset Photo Shop to
Develop Sheet Film in a Developing Tank**

This title states that a process report is being written, that the subject is film development, and that the process is limited to one process for one type of film. The words in the title are specific; unnecessary words are omitted.

- An informative report title can rarely be written in fewer than seven words; often it contains between fifteen and twenty words.

- Only specific, meaningful words are used in a title. Unnecessary expressions or clichés like those in the following example subdue the important information.

**EXAMPLES**

**A report on (concerning, about, dealing with, and similar prepositions)
A study of
An investigation of
A discussion of**

Unless a writer can justify the use of these or similar expressions in a title, they are unnecessary.

■ Small but important words like *a*, *an*, and *the* are included in report titles; telescopic style, a style that omits these and similar small words, is never used. For example, the telescopic title "Analysis to Determine Best Site for New Plant" *should* be revised to read "An Analysis to Determine the Best Site for a New Plant."

## Placing a Title in a Report

The length of a report determines whether a title is placed on the first page of text or on a separate title page. The title of a very short report (less than three pages long) is usually placed on the first page of text. The title begins 2 inches from the top of the page and is written entirely in capital letters.

Unlike titles for tables and formal classifications, the title for a report is double-spaced. Three line spaces separate the title from the text. It is horizontally centered on the page and indented at least five spaces from the left and right-hand margins of the text.

The title should be attractively arranged because it is the first item the reader sees. At least two lines should be used for a title. However, because of the information needed in most titles, three lines are most common. Four or five are acceptable. Lines of the title should vary in length, and when possible, line divisions occur where a logical break in text exists. Balance is also important. Approximations of the line arrangements shown in Figure 7–2 are recommended. Similar line balance should be used for two-, four-, or five-line titles.

Most formal reports extend beyond three pages and are preceded by a separate title page. A title page must be neat, attractive, and informative. It must also harmonize with the conservative tone of formal reports. A writer may design an original format or duplicate the title page illustrated in Figure 7–3.

As shown in the figure, the title page has three major divisions:

1. The title of the report begins 2 inches from the top of the title page. The title format explained previously is recommended. The entire title is usually capitalized. However, if a middle line contains only words that are not capitalized in any title, that line may be written in lowercase letters, as shown in the following example.

EXAMPLE

**A PHYSICAL DESCRIPTION**

**of the**

**17–400 SAW BLADE**

**FIGURE 7-2** An example of line arrangements for classification, table, and report titles

2. The name and address of the person or company for whom the report is written is single-spaced and is typed in the center of the page. Introductory words such as *Prepared for* or *Submitted to* are written two spaces above the name. Only the first letter of each important word in this division of the page is capitalized.
3. The author's name and the date on which the report is submitted are arranged so that they do not extend below 1 inch from the bottom of the page. If the author's title and address are included in this division, they are placed on separate lines below the author's name. The information about the author is single-spaced. The date is placed either one or two spaces below the author information. Only the first letter of each important word is capitalized. The word *by* is placed two spaces above the author's name.

All parts of the title page are horizontally centered. The backing sheet shown in Figure 7-1 has center vertical and horizontal lines that simplify the preparation of a title page for writers whose typewriters do not center automatically.

**AN ANALYSIS CONCERNING THE POSSIBLE USE
OF NUCLEAR ENERGY IN MINING**

**Prepared for**

**John K. Allen
Project Engineer
Smith-Allen Company
Akron, Ohio**

**by**

**Bill L. Thompson
Production Foreman
September 29, 1986**

**FIGURE 7-3** An example of a technical report title page

# TOPIC HEADINGS IN REPORTS

Topic headings have three important functions in formal reports:

1. To inform the reader that one topic is completed and a new one is being introduced
2. To permit a reader to find one part of a report quickly
3. To enable a reader scanning the text to see the organization of information

Headings are used in all formal reports and should also be included in most memorandums and letter reports. Although many levels of headings can be discussed theoretically, a technical report rarely requires more than three:

1. A center heading
2. A margin heading
3. A paragraph heading

## Placing Headings in Reports

Formats for the three kinds of headings — center, margin, and paragraph — are shown in Figure 7-4. These formats are explained in the following paragraphs.

1. Center headings are used for major divisions, including the introduction, for each major topic discussed in the body of the report, and for the summary. These are the topics preceded by Roman numerals in an outline. Center headings, written entirely in capital letters, are horizontally centered on the page. Three spaces separate them from the text.
   The first major heading in the text is the word *introduction*. When a title page is used, *introduction*, written as a center heading, is placed 2 inches from the top of the first page of text. If a title page is not used and a title is written on the first page of text, the word *introduction* is usually omitted. Under these circumstances, no other heading can be used until the first major topic in the body of the text has been introduced.
2. Margin headings are first-level subtopics. In the outline, capital letters precede them (see Figure 6-1). "Definition of Terms" is a first-level subtopic. When this and other first-level subtopics are used as margin headings in a report, all important words are capitalized (see Figure 7-4). The entire heading, placed on the left-hand margin of the text, is underlined. Normal double spacing is used above and below the margin heading.
3. Paragraph headings, preceded by Arabic numerals in the outline, are

---

**INTRODUCTION**

A feasibility study completed June 6, 1986, shows that installing a computer network will increase efficiency and lower costs at PBC Company.

<u>Purpose of the Report</u>

In this report, the components of the computer network are briefly described. Then the process involved in its installation is explained in detail.

<u>Definition of Terms</u>

The words <u>computer network</u>, <u>host</u>, and <u>workstation</u> as they apply to this report are defined as follows:

<u>Computer Network.</u> A computer network is a system that connects several personal computers to a single mainframe computer, a printer, and a variety of disk and tape drives.

<u>Host.</u> A host is the combination of a mainframe computer, a printer, and the disk and tape drives that send information to and receive information from each personal computer.

<u>Workstation.</u> The workstation is a term synonymous with <u>personal computer</u> when the computer is connected to the host.

---

FIGURE 7-4 An example of a center heading, a margin heading, and a paragraph heading

second-level subtopics. In Figure 7-4, each technical term being defined is shown as a paragraph heading. Because paragraph headings introduce paragraphs, they are indented five spaces from the left-hand margin of the text. They are underlined and followed by a period. The first letter of each important word is capitalized. The paragraph text begins two spaces after the period. Normal double spacing is used above and below the text.

## Planning Effective Headings

Headings, like other aids in technical writing, can detract from the general format of the text if they do not conform to standards required for formal writing. The following guidelines are offered to help a technical writer use headings effectively.

- Headings are written as words or phrases, not as complete sentences.

- Headings equal to one another in importance are parallel to one another in structure. For example, if one center heading begins with a noun, all center headings in a specific report begin with nouns. A margin heading is not necessarily parallel in structure to a center heading, but all margin headings are parallel to one another. A violation of this rule is also a violation of the rule for parallel structure.

- At least one full sentence is written between any two headings. Preferably, one or more paragraphs are written.

- The first sentence following a heading cannot begin with a vague pronoun such as *this, that, these, those,* or *it.* For example, if "Power Amplifier" is a heading, the next sentence may state "A power amplifier is a term reserved for an amplifier that supplies appreciable powers to a load." The pronoun *it* or *this* cannot correctly replace *power amplifier* at the beginning of the sentence.

- Infrequent use of headings in a report lessens their value as guides for the reader. Preferably, at least one heading is used on each page of text. When graphic aids, classifications, or lists of any kind are included in a topic discussion, this suggestion cannot always be observed. A writer can anticipate how many headings are needed in a specific report if the headings are considered as aids for the reader.

- A center or margin heading placed near the bottom of a page should be followed by at least two full lines of typewritten text. A paragraph heading should be followed by at least one full line of text. This rule corresponds with the rule that states paragraphs should not be introduced on the last line of a page.

- A center heading is never followed by only one margin heading unless the margin heading is followed by two or more paragraph headings. Likewise, a margin heading cannot be subdivided into only one paragraph heading; there must be at least two paragraph headings.

# SUMMARY

1. Format is the physical appearance of a completed report. Because format can influence a reader's attitude toward a report, using an appropriate format is a basic principle of effective written communication.

2. Writers using sophisticated electronic systems can easily create attractive formats. However, a writer who uses these systems to impress rather than to express can create a format that draws attention to itself, distracting a reader from important report content.

3. Suggested formats for text, report titles, and topic headings are discussed in detail. Examples are also given. These suggestions are guidelines, not rigid rules. However, writers should use judgment before discarding them for others.

4. Formats for graphic aids are explained in Unit 5. Writers using these aids should coordinate the guidelines given there with those given in this unit.

UNIT **7**

# SUGGESTED ACTIVITIES

A. Make a backing sheet. Use a smooth white sheet of unlined 8 1/2-inch by 11-inch paper. Following Figure 7-1 as a guide, draw margin lines for a 1 1/2-inch left-hand margin, both a 1-inch and a 2-inch top margin, a 1-inch bottom margin, and a 1-inch right-hand margin. The lines must be dark enough to be seen easily through the typing paper to be used for your report. Determine whether you want to include vertical and horizontal centerlines to be used as guides for preparing a title page. The horizontal centerline is two spaces above center because center material on a title page is more attractive when it is placed slightly above center.

B. Design a title page. Select a technical subject with which you are familiar. Plan a title that meets the requirements for report titles, and determine how much information your title page requires. Arrange the material neatly and attractively on a white 8 1/2-inch by 11-inch sheet of typing paper.

C. Using the report studied in Suggested Activities, Part B, Unit 6, or another of your choice, compare the text, title, and heading formats used on that report with the formats recommended in this text. Determine which formats you prefer and be prepared to defend your choice in a classroom discussion.

=== SECTION

# SUGGESTED SUPPLEMENTARY READING ===

Damerst, William A. *Clear Technical Reports.* New York: Harcourt Brace Jovanovich, Inc., 1972. In "Definition," pages 99 to 107, specific, detailed aids for writing definitions are given. Distinguishing characteristics between classification and partition are explained on pages 107 and 108.

Houp, Kenneth W., and Thomas E. Pearsal. *Reporting Technical Information.* New York: Macmillan Publishing Company, Inc., 1984. The authors discuss headings on pages 228 to 229. An example of a sentence outline is shown on page 162. Format is explained on pages 160 to 164.

Lannon, John M. *Technical Writing.* Boston: Little, Brown and Company, 1985. The author discusses definition on pages 111 to 127, partition on pages 135 to 137, classification on pages 139 to 144, outlining on pages 151 to 161, and format on pages 232 to 238.

Markel, Michael H. *Technical Writing: Situations and Strategies.* New York: St. Martin's Press, Inc., 1984. The author discusses outlining on pages 70 to 80, definitions on pages 131 to 139, classification and partition on pages 147 to 153, headings on pages 106 to 109, and title pages on pages 235 to 238.

Rathbone, Robert R. *Communicating Technical Information.* Reading, Mass.: Addison-Wesley Publishing Company, Inc., 1966. "The Random Order," pages 71 to 81, discusses the major parts of a report and techniques for effectively organizing the report.

Roundy, Nancy, with David Mair. *Strategies for Technical Communication.* Boston: Little, Brown and Company, 1985. Definitions are discussed on pages 121 to 135.

Sherman, Theodore A., and Simon S. Johnson. *Modern Technical Writing.* Chapter 4: "Organizing" (outlining). Englewood Cliffs, N.J.: Prentice-Hall, Inc., 1983.

The following textbooks give detailed information, guidelines, and illustrations for preparing graphic aids.

Damerst, William A. *Clear Technical Reports.* "Graphic Aids," pages 128 to 140. New York: Harcourt Brace Jovanovich, Inc., 1972.

Houp, Kenneth W., and Thomas E. Pearsal. *Reporting Technical Information.* Chapter 12: "Graphic Elements." New York: Macmillan Publishing Company, Inc., 1984.

Lannon, John M. *Technical Writing.* Chapter 13: "Visuals." Boston: Little, Brown and Company, 1985.

MacGregor, A. J. *Graphics Simplified.* Toronto: University of Toronto Press, 1981.

Only 58 pages long, this book teaches simple techniques for preparing graphic aids and for choosing appropriate graphic aids.

Markel, Michael H. *Technical Writing: Situations and Strategies*. Chapter 10: "Graphic Aids." New York: St. Martin's Press, Inc., 1984.

Roundy, Nancy, with David Mair. *Strategies for Technical Communication*. Chapter 15: "Visual Aids." Boston: Little, Brown and Company, 1985.

# Writing
# The Report

# The Descriptive Report

After reading this unit, you should be able to do the following tasks:

- Use description in a technical report.
- Organize a descriptive report.
- Write a descriptive report.
- Proofread a descriptive report.

## OVERVIEW

As a result of increasingly rapid developments in all businesses and industries, few employees can escape the need to prepare or help prepare written messages. Section III discusses and illustrates six types of formal reports and various informal reports. The kind of work an employee performs — as a researcher, an investigator, a machinist, a field engineer, and so forth — determines the kind or kinds of messages he or she writes. This unit explains and illustrates a physical description. Other kinds of reports are discussed in Units 9 to 15.

A descriptive report, as the term is used in this unit, details the physical characteristics of a machine, a device, or other equipment. In addition, it explains the relationship of one part of the mechanism to all other parts so that the reader can visualize the mechanism as a unit.

# USING DESCRIPTION

A description may be short enough to be part of an expanded definition, as shown in Unit 3. It may be combined with a process, an analysis, or an investigation. For example, the materials and equipment would probably be briefly described in a report explaining a process for growing commercial yeast. Often, however, a physical description is written as a complete report, as shown in Figures 8-1 and 8-2.

# ORGANIZING THE DESCRIPTIVE REPORT

Logical organization of a descriptive report is essential. Unless the development of one major part of a mechanism leads the writer and the reader logically to the second major part, the description lacks unity. Thus, the purpose for writing the report, to help the reader understand the mechanism, is not accomplished. Figure 8-1, an outline for a descriptive report, shows how a report describing the parts of an experimental yeast fermentor was organized.

An outline for a descriptive report shows all the information needed in the report and is prepared before the report is written. A complete outline like the one in Figure 8-1 is essential for descriptive reports that must be detailed and specific. The following guidelines were used to organize the outline in Figure 8-1.

- The outline introduction always includes background information and purpose.

- Omitted data in a descriptive report is usually due to careless observation and cannot be justified. Therefore, the omitted-data section of the basic outline explained in Unit 6 is seldom included.

- Sources of data are included only if the technician obtains information from sources other than observation of the mechanism being described.

- Definitions are included if the report contains technical terms or mechanical parts that may be unfamiliar to the reader. All potential readers of the report in Figure 8-2 would be familiar with the technical terms used. No definition is needed.

- The major parts of the mechanism are introduced as the major topics to be developed in the report. The writer should observe the mechanism being described while outlining the report to see the relationship between parts and to see the physical detail of each part.

- The arrangement of major topics in the outline is determined by the relationship of one part of the mechanism to another. For example, the

I. Introduction
    A. Background
    B. Purpose of the report
    C. Major topics
        1. The container
        2. The agitator assembly
        3. The drive magnet
II. The container
    A. Material
    B. Shape
    C. Dimensions
    D. Cover
III. The agitator assembly
    A. The agitator shaft
        1. Material
        2. Dimensions
    B. The agitators
        1. Use
        2. Material
        3. Connection to the shaft
        4. Location
        5. Components
            a. Collar and plate
            b. Impellers
                (1) Type
                (2) Location
                (3) Size
    C. The air sparger
        1. Location
        2. Size
    D. The driven magnet
        1. Location
        2. Dimensions
IV. The drive magnet
    A. Location
    B. Use
V. Summary

FIGURE 8-1 An example of an outline for a descriptive report

drive magnet listed in Figure 8-1 is the last major topic because its function is understood only after other parts of the yeast fermentor have been described.

- When a major part of a mechanism is made up of minor parts, subdivisions are used. For example, the agitator assembly is subdivided into four parts.

- The physical characteristics such as size, shape, color, and material of each mechanical part introduced into the outline are listed as subtopics under the part to which they belong. See the detailed development of the three major parts of the fermentor.

- The use of a mechanical part is not a physical characteristic. Therefore, use is included in descriptive reports only when a brief explanation of use clarifies the relationship of one mechanical part to another. For example, in the outline of Figure 8-1, the use of the drive magnet is necessary because it justifies that magnet as part of the fermentor.

- The summary of a descriptive report is a condensed version of the information given in the report. Conclusions and recommendations are rarely needed.

## WRITING THE DESCRIPTIVE REPORT

A descriptive report developed from the outline in Figure 8-1 is illustrated in Figure 8-2.

---

**THE PHYSICAL DESCRIPTION
OF THE EXPERIMENTAL YEAST FERMENTOR
USED FOR RESEARCH
AT BIOCHEMICAL ENGINEERING LABORATORIES**

*Background*

Walton Products Incorporated has asked Biochemical Engineering Laboratories to study the effect of motion and/ or ingredient mix on yeast-cell growth. The company plans to use the data from this study to design a reactor system that will produce genetically altered yeast cells. For this project, the Biochemical Engineering Laboratories has purchased a 2-liter experimental yeast fermentor that has a working capacity of 1 1/2 liters.

---

**FIGURE 8-2** An example of a report showing the development of a physical descriptive report and the relation of a report to an outline

*Purpose to be
accomplished
and major
topics listed*

**The physical characteristics of the three major parts of
this fermentor -- the container, the agitator assembly, and
the drive magnet -- are described in detail in this report.
The information given also shows the relationship between
the stirring mechanism and the drive magnet.**

**THE CONTAINER**

**The experimental yeast fermentor being described is
shown in Figure 1.**

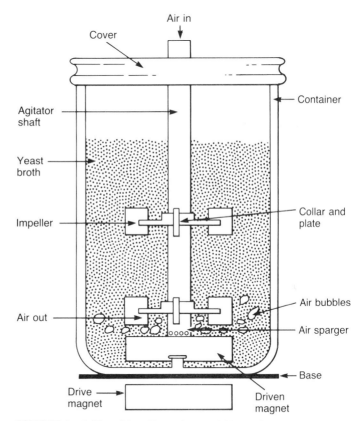

**FIGURE 1.  A Two-Liter Experimental Yeast Fermentor**

*Detailed de-
scription of the
container*

**The container shown is a Pyrex glass, cylindrically shaped
jar. It measures 9 inches in height and 5 inches across the
outside diameter at the top.  At the bottom, it curves to a
4-inch diameter.  When the fermentor is in use, a rubber
headplate is secured to the top of the jar with a retaining**

**FIGURE 8-2** *(Continued)*

*Lead-in sentence*

ring. In the center of the jar is the agitator assembly. See Figure 1.

## THE AGITATOR ASSEMBLY

*Introduction of minor parts*

The agitator assembly is made up of four parts: an agitator shaft, two agitators, an air sparger, and a driven magnet.

### The Agitator Shaft

*Detailed description of first minor part*

*Lead-in sentence*

The agitator shaft is a stainless-steel pipe that extends from 1/2 inch above the jar into a magnet (described later) 1 1/2 inches above the bottom of the jar. The shaft measures 8 1/2 inches in length. It has a 1/2-inch outside diameter and a 3/8-inch inside diameter. Two agitators encircle this shaft.

### The Agitators

*Detailed description of second minor part*

The agitators shown in Figure 1 are used to mix the yeast broth. Both are constructed from stainless steel. Small screws connect them to the shaft. One agitator is 5 inches from the top of the shaft; the other is 2 1/2 inches lower.

*Topic sentence*

Each agitator has two components: a cast collar and plate that encircles the agitator shaft and four turbine-blade impellers. The impellers are attached equidistantly around the plate circumference. Each impeller is 1/2 inch square and 1/16 inch thick. The entire agitator has a 2-inch outside diameter. Below the lower agitator is the air sparger.

### The Air Sparger

*Detailed description of third minor part*

The air sparger consists of ten perforations in the agitator shaft. Placed at equal distances around the shaft circumference, each perforation measures 1/16 inch in diameter, as shown in Figure 1.

### The Driven Magnet

*Detailed description of fourth minor part*

*Lead-in sentence*

The agitator shaft extends into a driven magnet (see Figure 1) immediately below the sparger. This magnet measures 2 inches in diameter and 1/2 inch in thickness. The entire agitator assembly is activated by a drive magnet.

**FIGURE 8–2** *(Continued)*

**THE DRIVE MAGNET**

*Detailed de-scription of the drive magnet and its relation-ship to the agitator assembly*

The container and the agitator assembly are secured to a stainless-steel base in which a drive magnet is enclosed. This magnet has dimensions identical to those of the driven magnet. The drive magnet, when attached to a solid-state electronic controller, sets the agitator assembly in motion through an interacting magnetic field.

**SUMMARY**

*Recapitulation of report text*

The experimental yeast fermentor purchased by Bio-chemical Engineering Laboratories for the Walton Products Incorporated project consists of a container, an agitator as-sembly, and a drive magnet. Each of these parts has been illustrated and physically described. This 2-liter mechanism seems adequate for the required yeast-growth studies.

**FIGURE 8-2** *(Concluded)*

Although each topic is listed in the outline as a self-contained unit of information, one topic cannot be isolated from another topic in the report. Writing skill and the use of techniques including transitional words and phrases, lead-in sentences, and headings are needed to make the report function as a unit of interrelated facts. Some useful techniques for developing a descriptive report from an outline are explained here and illustrated in Figure 8-2.

1. A review of the completed outline helps a writer visualize the entire report, not just individual divisions. This review helps the writer deter-mine the effective use of headings, graphic aids, transitional aids, and lead-in sentences.
2. A report is never limited or extended to fill a specified number of pages or to use a specified number of words. Instead, the writer works to achieve the purpose of technical writing, which is to transmit pertinent informa-tion to a reader.
3. The purpose of the report, which develops logically from the background information, must be stated specifically, clearly, and completely. The purpose section in Figure 8-2 states that a descriptive report is being written and that the reader can expect a detailed explanation of each part of a specific mechanism.

4.  In the body of the report, every physical characteristic of each mechanical part introduced into the report is described. Figure 8–2 illustrates a detailed description of the three parts of a yeast fermentor: the container, the agitator assembly, and the drive magnet.

5.  Center headings are used in all reports. Margin headings and paragraph headings are used when the information is extensive enough to justify them. In Figure 8–2, no paragraph headings are used. Margin headings are used only to introduce the four parts of the agitator assembly.

6.  Although headings are aids to understanding, they are not substitutes for necessary transitional words and phrases or lead-in sentences. For example, a lead-in sentence usually introduces each major topic (center heading) used in the body of a report.

7.  Graphic aids are needed in all reports that describe the physical characteristics of a mechanism. The graphic aids, however, are not substitutes for words; they are used to intensify the verbal explanation.

8.  Graphic aids chosen for descriptive reports are usually drawings or photographs. They should be selected carefully so that the ones used show all the physical details discussed in the report. (See Figure 1 within the report.)

9.  Determining how many graphic aids to use can be difficult. In Figure 8–2, only one graphic aid, a drawing showing the entire fermentor, was chosen for three reasons: (1) the fermentor being described is a simple mechanism, (2) the single drawing shows the relationship of one part of the fermentor to another part, and (3) the text is short; therefore, three or four graphic aids would overshadow written information.

10. The summary of the descriptive report renames the major parts of the mechanism being described and any important relationship between one part and another. It also explains how the discussion has accomplished the objective stated in the purpose of the report, as illustrated in Figure 8–2. A summary never introduces new information.

## PROOFREADING THE DESCRIPTIVE REPORT

A writer can rarely write a first draft of a report that cannot be improved. Therefore, careful proofreading of all reports is essential. A beginning writer who wants to make his or her reports reflect careful planning, effective presentation of information, and consideration for the reader reviews the first draft of the report at least three times.

In the first review, the writer studies the entire text to check for completeness, clarity, accuracy of information, and accomplishment of the purpose stated in the introduction. The writer also notices whether the text has logical development and continuity and whether the headings and graphic aids are correct.

In the second review, each major topic is checked as the writer considers the following questions:

—Is all information specific? If any approximations have been used, can they be justified? Does the reader understand why the approximation replaces specific data? Are specific names given to all parts of the mechanism?
—Are all words spelled correctly?
—Is punctuation correct and meaningful to the reader?
—Are sentences carefully constructed so that faulty parallelism, dangling modifiers, vague antecedents, and misplaced modifiers are avoided?
—Do any sentences begin with *there* or the expletive *it*? If so, can the sentence be rearranged?
—Do the words used in the text say what the writer wants them to say? Can short, familiar words replace any long or unfamiliar terms used in the text? Are unfamiliar technical terms defined in enough detail to permit the reader to understand them?
—Can explicit words replace any long, involved phrases?
—Has the rule for consistency in the use of numbers been applied?
—Have any abbreviations or symbols been used in the text? If so, can they be justified?
—Have rules for capitalization been applied to the text, headings, and titles?
—Have hyphens been used for all words that require them?

In the third review, the writer again reads through the complete text to make sure that the report meets the following five standards required for technical writing.

1. The meaning is clear.
2. The writing is concise.
3. The facts are correct.
4. The information is complete.
5. The report is written in standard English.

# SUMMARY

1. Reports are needed in nearly all organizations. The kinds and lengths of reports used can vary extensively.

2. A person involved in designing, constructing, or using a mechanism or device is often required to give its physical description in a report. Someone involved in projects such as land development, geological mining, or archeological studies may also write physical descriptions.

3. Physical description can be an important major topic in other types of reports: processes, analysis, trouble-shooting, or examination (Units 9 through 12). When used in one of these types, it is usually the first topic discussed.

4. Physical descriptions are most easily and accurately written if the writer can fill in an outline and take notes while observing the item being described.

5. Other information such as a use or a process is part of a description only when this information contributes to a reader's understanding of the item being described.

6. Graphic aids are essential parts of a description. An effective writer refers to them frequently as he or she develops the text.

7. A physical description must be organized to show how one part relates to another. Thus, by the end of the report, all parts are systematically integrated.

8. Proofreading while the writer visualizes his or her reader's needs is an essential part of writing a useful physical description.

UNIT 8

# SUGGESTED ACTIVITIES

A. Select a simple mechanism that functions as a unit but that consists of two or more major parts. While observing the mechanism, prepare a detailed outline. Begin by establishing the title and the purpose for the report to be written later. Do not copy the introductory topics used in Figure 6–1. For example, use "Source of Data" if it is needed. Use "Definition" if definitions are required. If any approximations must be used in the report, justify them in the part of the introduction reserved for omitted data. Review the outline as a unit of detailed information before writing the report.

B. Following the outline, write a descriptive report.

C. Using the suggestions given in this unit, proofread the report. (Some weaknesses in the text may be more easily recognized if the report is read aloud than if it is read silently.)

D. Ask a person, preferably one unfamiliar with the mechanism being described, to give constructive criticism of the report.

E. Determine whether a title page or a first-page title is to be used; then write the final draft of the report in accordance with the guidelines for format explained in Unit 7.

# The Process Report

## OBJECTIVES

After reading this unit, you should be able to do the following tasks:

- Describe the process report and its purpose.
- Organize a general or detailed process report.
- Write a process report.
- Determine a consistent point of view for a report.
- Proofread a process report.

## OVERVIEW

A process is a method of doing something that involves a series of operations. A process report, therefore, explains how a product is produced, a test is completed, or a machine is operated by including the details of procedures used to perform a series of operations. Processes are a basic part of nearly every industry. Consequently, many technicians have the responsibility of writing process reports. Because industries and processes performed within each industry are diversified, process reports differ from one another in length, content, and method of organization and development. This unit shows how a process report may be general or detailed. It then illustrates and explains a detailed report.

# ANALYZING PROCESS REPORTS

A process report may be general, written for a person or persons not directly involved in performing the process. For example, general explanations of processes may be written for a magazine, newspaper, or similar publication. Even within a company, general explanations of processes can help employees relate their work to the processes developed by other employees. For example, machinists may need a general understanding of work to be performed by the equipment they machine. Members of management, who read many reports, may prefer a general explanation of some processes to a detailed report.

Another process report, like a descriptive report, can be detailed. A detailed report is designed to give the reader an insight into all technical procedures needed to complete the process. The type of reader for whom this kind of report is written varies widely — for example, a new employee, a supervisor, a member of management, a salesperson, or anyone else who can understand and use detailed information.

As stated previously, a technical process report may explain how a machine functions, how a test was conducted, or how a new product was developed. Because process reports vary extensively in detail and subject matter, a flexible plan for developing them is needed. Therefore, in addition to studying the information given in this text, a writer preparing a process report should analyze the subject, the reader, and the purpose of the report to determine the following:

1. How much information should be included
2. Whether the information should be specific or general
3. How the information can be logically organized and developed

# ORGANIZING THE PROCESS REPORT

The purpose to be accomplished in the process report is the basis for organization. If the report is general, the amount of detail included in the outline differs from the amount used for a detailed one. If the report explains the operation of equipment, the arrangement differs from that used for a report explaining only a procedure. Finally, if the report is written for a reader who understands the process, the organization and content differ from that written for one who has little or no knowledge of the process.

## Organizing the General Process Report

Before a general process report is written, the writer determines who the reader is, why the reader wants the report, and how much detail is required.

Beginning technicians studying a process related to their work need a technically detailed explanation. Laypeople seeking only superficial knowledge, however, are often bored and confused if the writer presents technical data.

The general process report is seldom difficult to organize. For instance, in four major sections, the writer can explain, in layperson language, the manufacture of bits used for hard-rock drilling:

1. The introduction to the report includes background information and the purpose to be accomplished. The introduction should also have a "Definition" subsection for defining the terms *alloy, matrix powder, infiltration, drill-bit shank, drill-bit crown,* and *crown mold.* If the text needed to define these or other terms used in a general report seems excessive, the definitions may be given in a glossary (see Unit 16). The "Definition" section, referring the reader to the glossary, is still used. The final subsection in the introduction presents the major topics of making the drill-bit shank and making the drill-bit crown.
2. A nontechnical discussion of cutting, shaping, and threading the shank completes the explanation for making the drill-bit shank.
3. A nontechnical discussion of selecting the crown mold, filling the mold, preparing the cylindrical funnel, infiltrating the matrix powder, and heating the mold completes the explanation for making the crown.
4. A short summary completes the report.

Each part of the discussion is written in nontechnical language, easily understood by a layperson. Graphic aids, nearly always used in process reports, are incorporated if they contribute to the information, interest the reader, and help the reader understand the report.

## Organizing the Detailed Process Report

Routinely, employees involved in projects must write detailed process reports, not only to inform a reader but also to record how something was done. For example, a process report may be written when a manufacturer improves an old product or develops a new one, when a social service employee works with a client, when a technician in a laboratory conducts tests, or when a technician installs equipment.

A detailed process report, especially one used as a record, includes all technical data needed to permit duplication of the process. The writer, therefore, focuses upon specific detail and logical organization — factors that are more likely accomplished if the writer keeps notes while performing the process. Figure 9–1 illustrates one possible development for a process report.

Planning the detailed process report can involve problems only the writer can solve. For example, the writer may have to consider the following questions:

—Are either materials, equipment, or both used in the process?
—Are the same materials used for each process step in the report?
—Does one major item, such as the door that is the subject of the outline in Figure 9-1, require physical description if the process is to be understood?
—Will the amount of technical data needed to verify the process distract the reader from the process?
—Should some of the data be placed in the appendix and referred to in the report?

I.  **Introduction**
   A. **Background**
   B. **Purpose**
   C. **Major topics**
      1. **Physical description**
      2. **Test preparations**
      3. **Fire test for the door**
      4. **Hose-stream test for the door**
      5. **Monitor of the test procedures**

II. **Physical description**
   A. **Fireguard door design**
   B. **Construction of the door**
      1. **Panels**
         a. **Construction**
         b. **Interconnection**
      2. **Fire liner**
         a. **Attachment**
         b. **Location**
         c. **Design**
      3. **Tracks**
         a. **Number**
         b. **Location**
         c. **Use**
      4. **Lead post**
   C. **Door operation**
      1. **Method**
      2. **Open position**
      3. **Closed position**

**FIGURE 9-1** An example of a topic outline for a report explaining the process used to test Won-Door's FireGuard Door

III. **Preparations for testing the door**
    A. Positioning the door
    B. Preparing the furnace
       1. Spacing the thermocouple rods
       2. Spacing the burners

IV. **Fire test for the door**
    A. Increasing the heat intensity
    B. Monitoring the thermocouple rods

V. **Hose-stream test for the door**
    A. Purpose for testing
    B. Equipment used
    C. Application of the hose stream
       1. Initial hose location
       2. Hose movement
       3. Duration of the test

VI. **Study of the test data**
    A. Thermocouple and thermocouple rod temperatures
    B. Condition of the face side
    C. Condition of the back-face side

VII. **Summary**

**FIGURE 9-1** *(Concluded)*

Only a writer who understands the report subject and is aware of his or her readers' needs can answer these and similar questions. The answers can significantly affect the quality of the report. They should be based upon the following guidelines:

- A listing or description of materials or equipment has no value if it does not precede the process step involving it. For this reason, a listing or description is often the first major topic of the report; see the physical description section in the outline of Figure 9–1. A short description or list used in only one process step may introduce that step. See the second paragraph in "Hose-Stream Test for the Door" in Figure 9–2 given later in this unit.

- The discussion of the process should not be interrupted periodically while the writer introduces equipment or materials.

| EXAMPLE | **The filament switch is turned to "On." (The switch is located on the left-hand side of the control box.)** |

Interruptions of this kind are distracting to the reader, who expects to move systematically through a step-by-step procedure.

- Data placed in an appendix has no value unless the reader has been told it is there and the writer has explained, preferably in the report introduction, its relationship to the report subject. Other references to the appended material may be needed within the report text.

The preceding guidelines relating to equipment and data helped the writer develop the outline shown in Figure 9–1. The general guidelines that follow were also used.

- No formal or expanded definition or omitted-data reference is needed. Therefore, the report introduction includes only background, purpose to be accomplished, and an introduction of major topics.

- The physical description of the Won-Door FireGuard door as the first major topic relates the process to one particular type of fire barrier. It also prevents frequent interruptions once the writer begins to explain the process.

- The process steps are presented in the order in which they occurred.

- A short summary completes the outline.

# WRITING THE PROCESS REPORT

Figure 9–2 illustrates the report developed from the outline shown in Figure 9–1. The following comments explain how the report was developed and why this method of development was used.

1. The report was written primarily as a reference source for any person interested in protection from fire or in the potential safety of various fire barriers.
2. The short introduction adequately introduces any potential reader to the reason for testing (the report background), the purpose to be accomplished, and the four major steps to be discussed and illustrated.
3. A title page was prepared for this report, partly because of its potentially varied readership. The following example shows title and title format used.

<table>
<tr><td>

━━━━━━━━━
**EXAMPLE**
━━━━━━━━━

</td><td>

**THE TESTING PROCEDURE**
**UNDERWRITERS LABORATORIES INC. USED**
**ON A WON-DOOR FIREGUARD DOOR**

</td></tr>
</table>

The title does not designate whether the report is general or detailed. A writer who considers this designation important includes it in the statement of purpose.

4. Two graphic aids are not commonly used in a short description. However, a single aid could not adequately illustrate important features of the door. The other graphic aids were also selected to support the text.
5. Only four margin headings are used in the report. Each heading indicates a significant change from one phase of a major process step to another. Although an outline helps a writer select appropriate headings for a report, every part of an outline is not included as a heading unless the discussion of each outline listing is one or more paragraphs long and involves a major shift in subject matter.
6. Only a restatement of information and the significance of the test are needed in the summary. However, some processes result in conclusions and recommendations. If, for example, a new process were being explained and the writer believed the process should be adopted, conclusions and recommendations, discussed in "Developing the Summary" (Unit 6), would be needed.

---

**INTRODUCTION**

*Background*

Won-Door Corporation, Salt Lake City, Utah, needed certification that its unique FireGuard door met the American Society for Testing Materials (ASTM) E152 and the Underwriters Laboratories Inc. (UL) 10-B test for fire doors. This test determines the amount of time a particular fire door can protect an area from fire. Code officials use the test as a basis for determining whether a door meets the requirements specified in the national building codes.

UL representatives inspected the door in Salt Lake City and approved it for testing. Therefore, the test door was shipped to UL facilities in Northbrook, Illinois, where the tests would be conducted.

**FIGURE 9-2** An example of a process report (Source: Jay Smart, board chairman/CEO, and Ron Smart, president, Won-Door Corporation, Salt Lake City, Utah)

*Purpose and major topics combined*

The text and graphic aids in this report describe the physical characteristics of this door and then explain the testing process. The test involved four procedures: (1) preparations for testing the door, (2) the fire test for the door, (3) the hose-stream test for the door, and (4) a study of the test data.

## DESCRIPTION OF THE WON-DOOR TEST DOOR

*Topic sentence*

The Won-Door FireGuard door actually consists of two accordion-fold doors, as shown in Figure 1.

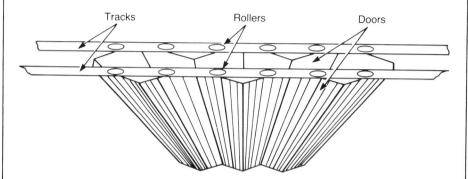

FIGURE 1.  A Drawing Showing the Two-Track, Two-Door System on Won-Door's FireGuard Door

SOURCE:  Eric Stilsen, Stilsen & Stilsen, Salt Lake City, Utah.

### Construction of the Door

*Lead-in sentence*

Each of the two doors is made from a series of 24-gauge steel panels interconnected with formed-steel hinges. Each panel is finished with bonded baked enamel and is, therefore, available in a choice of colors. A specially designed patented clip system holds a fire liner to the panels. This liner extends from the ceiling to the floor, preventing fire or smoke from filtering through the floor or ceiling area. The principle of the two-door system and the liner prevents the transfer of heat from the fire-exposed side to the unexposed side of the door assembly. A steel pin-and-

**FIGURE 9-2** *(Continued)*

roller system is used to suspend the panels from two separate, stationary, overhead tracks.  The tracks are spaced 8 inches apart (Figure 1).

### Operation of the Door

In the usual open position, the door recedes into a concealed wall area.  In the closed position, achieved automatically by an electronic closing assembly, it blocks all smoke from entering any area protected by FireGuard doors.  However, anyone trapped behind the closed door can escape by touching the exit panel at the lead post, as shown in Figure 2.

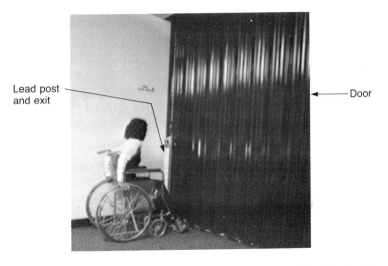

Lead post and exit

Door

FIGURE 2.  A Photograph Showing the Exit Panel at the Lead Post of the Won-Door FireGuard Door

SOURCE:  Won-Door Corporation, Salt Lake City, Utah.

The door automatically opens a predetermined distance to allow passage through; it then automatically closes again and seals.

*Lead-in sentence*

The Won-Door FireGuard door used for testing at UL measured 12 feet by 14 feet.  It weighed 5.2 pounds per square foot.

FIGURE 9-2 *(Continued)*

## PREPARATIONS FOR TESTING THE DOOR

The door arrived at UL in packed sections that allowed for easy installation according to specifications. Once again UL personnel inspected the door and positioned it for the test.

<u>Positioning the Door</u>

UL technicians placed the test door in its closed position (see Figure 3) in a heatproof masonry frame, which is part of the test furnace. The method of placement allowed only one side of the door, referred to as the face side, to be exposed to the furnace flames. The other side, the back-face side, was left unexposed and available for continual observation during the test.

Test door

Masonry →

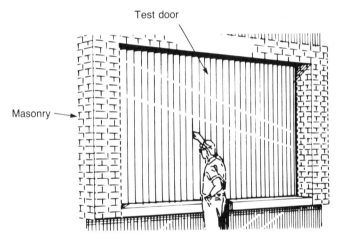

FIGURE 3.  A Drawing of the FireGuard Door Ready for Testing

SOURCE: Eric Stilsen, Stilsen & Stilsen, Salt Lake City, Utah.

<u>Positioning the Burners and Thermocouples</u>

*Lead-in*
*sentence*

The furnace consisted primarily of eighty-four gas-fueled jet burners directed toward the door; see Figure 4. These burners were arranged so that heat flames from them would give relatively even heat across the entire door surface. Thermocouples encased in sealed, heat-protecting ceramic tubes known as rods were placed among the burners (Fig-

FIGURE 9-2 *(Continued)*

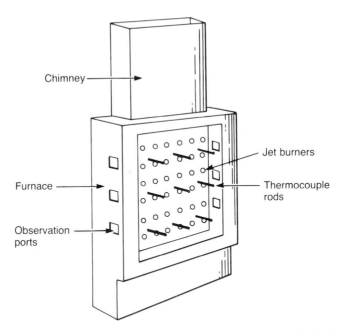

**FIGURE 4.   A Drawing Showing the Thermocouple-Rod and Jet-Burner Arrangement**

ure 4). They were positioned so that the thermocouple junctions would remain 6 inches from the test assembly during the entire firing process. They were also symmetrically spaced and distributed to show the temperature near all parts of the door.

In addition to these thermocouple rods, twelve thermocouples, one for every 14-square-foot area, were spaced uniformly on the back face of the door. These thermocouples would determine how much heat penetrated the entire test assembly. The door and furnace were now ready for the fire test.

### THE FIRE TEST FOR THE DOOR

At the beginning of the fire test, no heat was applied. The heat was then turned on. The flame intensity was steadily increased until flame from all eighty-four burners

FIGURE 9-2 *(Continued)*

exploded simultaneously and continuously against the entire face side of the door.

During the first 2 hours of the fire test, the temperature on the thermocouple rods was read every 5 minutes. The fire test continued for an additional hour. During this period, the thermocouple-rod temperatures were read every 10 minutes. During the entire test, both sides of the door were visually monitored, the face side through observation windows in the furnace (Figure 4). At the end of 3 hours, the furnace was shut off and the hose-stream test was begun.

### THE HOSE-STREAM TEST FOR THE DOOR

The hose-stream test is an essential part of fire-protection testing. It determines whether a fire barrier is uniformly structured. Therefore, immediately following the fire test, a hose stream was applied to the door.

The hose stream was delivered through a 2 1/2-inch-diameter fire hose equipped with a water-dispensing nozzle. This nozzle, 2 1/2 inches at the base, tapered to a 1 1/8-inch discharge tip.

As the test began, the tip of the nozzle was located 20 feet from the test door in line with its center. At the base of the nozzle, the water pressure was 45 pounds per square inch. From the center, the hose was moved slowly and regularly to every square inch of the door. This hose-stream test continued approximately 8 minutes.

Again, while this test was being conducted, the test assembly was constantly monitored. At the end of the test, the back face was scrutinized. Any water seepage through any area of this surface would mean that the door had failed to meet required standards.

*Lead-in and topic sentence*

### STUDY OF THE TEST DATA

After the tests were completed, all the computed and observed data were studied. From the study, the following facts were established:

- After 1 1/2 hours, the thermocouple rods on the face

**FIGURE 9-2** *(Continued)*

side of the door registered 1790 degrees Fahrenheit. At the end of the 3-hour fire test, the heat was 1980 degrees Fahrenheit. In contrast, the back-face temperature was always under 500 degrees Fahrenheit.

• The face side of the door showed considerable damage; some of it had disintegrated. However, the back-face side showed no damage of any kind. In addition, no water had seeped through any part of the back face, including the hinged areas.

• The low temperature on the back face indicated that the door exceeded all test criteria for fire protection, proving that the door effectively blocks the spread of fire into any area protected by it.

### SUMMARY

*Recapitulation*

UL personnel in Northbrook, Illinois, conducted fire and hose-stream tests on a Won-Door FireGuard accordion-fold door. A description of the door has been given and the processes used for testing have been explained and illustrated. The study of data taken during the fire and hose-stream tests has verified that Won-Door FireGuard doors excel as fire barriers; therefore, the door can now be used in areas not previously approved for accordion-type doors.

**FIGURE 9-2** *(Concluded)*

A report explaining fire tests for all fire barriers and written by fire-test personnel will differ significantly from that shown in Figure 9-2. A list of specifications will probably be required. Equipment and materials may replace physical description. Graphic aids such as charts and tables related to testing procedures and results will likely replace drawings and photographs.

The foregoing explanation shows that a single process can be perceived from more than one point of view. It thus demonstrates that the reader and the purpose to be accomplished determine the organization and the scope of all process reports. It also shows that judgment plays an essential role in preparing these reports.

# DETERMINING THE POINT OF VIEW

Maintaining a consistent point of view is often difficult for the person writing a process report.

In the following example, the first sentence is in the imperative mood (uses verbs that request or command), with the second-person viewpoint. The second sentence is in the indicative mood, with the third-person viewpoint.

---

**EXAMPLE**

**To measure the height of a standing tree, mark a distance of 100 feet from the base of the tree. The person measuring the height of the tree then stands on the point that has been marked.**

---

A change of moods is confusing to the reader and also suggests that a writer has not planned the report carefully before writing it. For this reason, a process report is always written objectively in the indicative mood.

In this respect, a process report must not be confused with a manual. The steps of the process in an instruction manual or a short instruction guide are written in the imperative mood, but the introduction to the text and any description of equipment and materials are written in the indicative mood. Therefore, to avoid a frequent change of moods, a writer preparing manuals or instruction guides should introduce equipment and materials before explaining the process. For example, if the process report in Figure 9–2 were written to give instructions, all process steps, including references to time, temperature, and so forth, would be written in the imperative mood. An explanatory heading preferably replaces a lead-in or introductory statement, as in the following example:

---

**EXAMPLE**

**APPLYING THE HOSE STREAM**

1. **Point the hose stream toward the center of the door.**

2. **Slowly and regularly move the stream across the door.**

3. **Continue to move the stream up, down, and across the door until every square inch has been hosed for 3 seconds.**

---

Carefully labeled graphic aids are required in all instruction manuals and guides that tell a reader how to perform a process. A summary is rarely used.

# PROOFREADING THE PROCESS REPORT

In proofreading the process report, the writer follows the same procedure as that used for proofreading the descriptive report outlined in Unit 8. The entire report is first checked for completeness, clarity, accuracy, and accomplishment of purpose. In addition, the writer should be able to answer "yes" to the following questions:

—Have the purpose and the reader been considered before the report is outlined?
—Has the indicative mood been used throughout the report?
—Are the equipment and materials needed for the process logically presented?
—Are the steps of the process arranged so that each step leads the reader logically to the completion of the total process?
—Are necessary graphic aids included? Are they identified by a figure number and a meaningful title? Are all components and processes shown in the graphic aids clearly identified?
—Does the report contain all details needed to satisfy the reader's needs?

Each section of the report is read and checked for weaknesses in sentence structure, grammar, punctuation, spelling, and correct use of numbers and capitalization. Then the complete report is reread so that the writer can be certain that the revised report is clear, concise, complete, and correct.

# SUMMARY

1. Process reports are written to explain a series of operations. These operations can relate to various numbers and kinds of activities extending from manufacturing, mining, testing, and investigating to social services.

2. Because scope, content, and point of view must be adapted to various kinds of process reports, this unit does not suggest specific guidelines for writing process reports. Identifying the purpose, the subject, and the reader and using logic and judgment usually provide the best guidelines.

3. Process reports are not "how to" documents. They tell what was done and how it was done. Therefore, the past and past perfect tense, indicative mood (not the imperative mood) are used.

4. An effective process-report writer develops a logical step-by-step explanation that allows the reader to follow the process progressively from beginning to end. He or she also avoids frequent interruptions to incorporate explanatory text such as location, definition, or description of equipment or materials.

5. Graphic aids can contribute significantly to a reader's understanding of an unfamiliar process.

6. Proofreading and, if necessary, revising a finished process report are always essential.

UNIT

# SUGGESTED ACTIVITIES

A. Select a process that can be completed in at least two but not more than four major steps. If possible, select a process that you have followed step by step and that you understand thoroughly. Take notes as you mentally or actually follow the procedure.

   Prepare a specific title that tells the kind of report being written and the limits of the text. Determine the purpose to be accomplished and the reader for whom the report is being written. Then develop an outline for a detailed explanation of the process.

B. From the outline, write a process report. Use meaningful headings and appropriate graphic aids.

C. Proofread the report. Then ask a person who is unfamiliar with the process to read the report and make constructive criticisms.

D. Determine whether to use a title page; then, using attractive format, write the final draft of the report.

E. Obtain an instruction guide or a process report from a technical library or another source. In a classroom discussion, analyze the text to determine whether an unfamiliar reader can easily follow the development of the process and understand all the words and technical terms used by the author.

# THE ANALYTICAL REPORT

## OBJECTIVES

After reading this unit, you should be able to do the following tasks:

- Describe the standards for analytical reports.
- Organize comparative analyses and single analyses.
- Write a comparative analysis.
- Proofread an analytical report.

## OVERVIEW

An analytical report evaluates a mechanism, a process, or a problem. An analytical report also answers questions concerning such issues as the kind of equipment needed for a particular job, the best location for a business, the need for hiring additional employees, and the feasibility of adopting new methods. These reports almost invariably require conclusions and recommendations because analyses establish the basis for decisions in nearly all industries, businesses, and professions.

In this unit, standards for analytical reports are established and methods for organizing both comparative and single analyses are discussed. Then, a comparative analysis is illustrated and explained.

# ESTABLISHING STANDARDS FOR ANALYTICAL REPORTS

Decisions based upon data presented in analytical reports are usually far-reaching and involve some risks. Therefore, the person who writes analytical reports should become familiar with the following standards:

1. Information used within the analysis must be complete. Recommendations based upon incomplete data can be erroneous.
2. Information must be scientifically researched, presented, and interpreted. Biased information leads to false conclusions. It is also unethical.
3. Information must be verified. A writer who states facts or figures leading to a conclusion and recommendation must defend all statements with references to reliable sources.
4. Information must be specific. General statements or approximate figures are not justifiable bases for recommendations. For example, a writer cannot recommend that a company purchase equipment or construct a building unless the statement of costs is exact and correct.

# ORGANIZING ANALYTICAL REPORTS

Specific kinds of analytical reports are too extensive to be discussed individually. However, analytical reports can be classified into two general kinds — comparative analysis and single analysis.

## Organizing Comparative Analyses

Comparative analysis is used to evaluate, using pertinent information, two or more alternatives. For example, comparative analysis may be used if a company is considering several sites in order to select the best one for a new building, if two or more plans for a bridge or a highway are under consideration, or if an employer needs to evaluate increased work loads to determine whether to pay overtime or to hire new employees.

Comparative analyses are a series of formal classifications (Unit 4, page 36) in which two or more plans, processes, problems, or pieces of equipment are classified according to two or more common bases. For example, in the comparative analysis of the two microscopes outlined in Figure 10-1, each microscope is analyzed according to five bases: magnification, detail, grain, sharpness, and cost. In most comparative analyses, as in this one, the major topics are the items being compared; the common bases are discussed as subtopics under each major topic.

Sometimes, as shown in Figure 10-1, preliminary procedures such as testing must take place before the bases can be compared. Nevertheless, formal

I. Introduction
   A. Background
   B. Purpose
   C. Definition
   D. Sources of Data
   E. Topics for Discussion
      1. A 35-millimeter camera attached to Microscope A
      2. A 35-millimeter camera attached to Microscope B
II. A 35-millimeter camera attached to Microscope A
   A. Camera-microscope arrangement
   B. Procedure for testing Microscope A
      1. Light arrangement
      2. Focusing procedure
      3. Method of exposure
   C. Results of the test
      1. Technical problems
      2. Magnification
      3. Detail
      4. Grain
      5. Sharpness
   D. Cost of Microscope A
III. A 35-millimeter camera attached to Microscope B
   A. Camera-microscope arrangement
   B. Procedure for testing Microscope B
      1. Light arrangement
      2. Focusing procedure
      3. Method of exposure
   C. Results of the test
      1. Magnification
      2. Detail
      3. Grain
      4. Sharpness
   D. Cost of Microscope B
IV. Summary
   A. Recapitulation
   B. Conclusions
   C. Recommendations

**FIGURE 10-1** An example of an outline for a comparative analysis

classifications are always the framework of comparative analysis. No single outline can present and solve all the problems that may develop in the preparation of analytical reports. Each situation that creates a reason for preparing an analysis is unique. For example, if several items such as five furnaces are being analyzed according to only two bases such as durability and cost, the writer uses the bases *durability* and *cost* as the major topics and discusses each furnace as subtopics.

These variations for developing an analysis show that the writer must select a plan that logically relates the comparative data, the conclusions, and the recommendations to the subject and purpose of the report. The following guidelines help a writer develop a logical plan.

- The outline uses five of the six possible sections of introduction explained in the basic outline in Unit 6. The section "Justification for Omitted Data" is sometimes used when unalterable circumstances prevent the writer from obtaining complete or specific data.

- Analyses of new equipment, processes, or ideas often require that definitions be included in the report introduction.

- Sources of data are rarely omitted in analytical reports. A writer who presents specific facts and figures as justifications for a recommendation must prove that these facts and figures are reliable. The proof is guaranteed when companies or individuals quoting the data are given as references.

- Using the camera-microscope arrangement for each of the two major topics in the outline permits the writer to analyze the data related to each topic. From a comparison of the two sets of data, conclusions are easily formulated.

- Nearly all analytical reports are written because a person or company needs a recommendation based upon the analysis of the major topics. The recommendation is an outgrowth of the report content summarized in a conclusion. Therefore, conclusions and recommendations are essential parts of most analytical reports.

## Organizing Single Analyses

A single analysis differs from a comparative analysis in only one way — one major topic is being evaluated according to two or more bases. The topic being analyzed becomes the title of the report, as indicated in the following example.

**EXAMPLE**

**AN ANALYSIS TO DETERMINE WHETHER**
**LAND DEVELOPMENT INCORPORATED SHOULD**
**BUILD A TRAILER CAMP AT CANYON EDGE**

Each basis upon which the report subject is analyzed is a major topic. For example, a report given the foregoing title is analyzed according to five bases:

1. Is a trailer camp needed in Canyon Edge?
2. Is land available?
3. What is the cost of construction?
4. What are the maintenance costs?
5. What is the anticipated income?

If the company has previously determined that a camp is needed and that land is available, only three topics are discussed.

Thus, a single analysis is, in reality, a partition rather than a classification (see Unit 4). Report content divides a single item — a trailer camp, for example — into parts that need investigation.

The introduction for the report includes background, purpose, sources of data, and major topics. It may also include definitions and omitted data. Conclusions are derived from the discussion of the topics; recommendations are derived from the conclusions.

# WRITING THE COMPARATIVE ANALYSIS

Figure 10-2 illustrates the report written from the outline in Figure 10-1. The analysis in Figure 10-2 is designed to illustrate the logical development of the outline and the importance of specific and verified information. The following comments concerning Figure 10-2 explain some problems that may develop while an analytical report is being prepared. A writer solves these problems by using good judgment and thought, not by rigid conformity to rules or examples.

1. A combination of situations may lead to the writing of an analytical report. Therefore, the background information may vary in length from one or two sentences to one or more paragraphs. Background information, however, is limited to the amount needed to justify the purpose of the report.
2. The purpose should state, in addition to the type of report and the bases for development, that recommendations are given. The wording of the purpose should make the reader aware that the recommendations are based upon information presented in the report.
3. All specific data in an analytical report must be justified, either through the writer's logical analysis or through references to authoritative sources. However, justifications should rarely interrupt the discussion in the body of the report. For this reason, "Sources of Data" may be a long and varied part of the introduction. "Sources of Data" often includes written sources of information (Unit 17) as well as names of people and companies that submitted facts or figures.
4. The major topics must be introduced before the discussion begins. Failure to tell the reader how an analytical report is being developed can cause confusion. In addition to listing the major topics, the writer should state the bases upon which the topics are discussed. Two or more obvious similarities between the

## INTRODUCTION

Equipment Manufacturing Company must purchase a microscope that, combined with a 35-millimeter camera, can be used to photograph surface finishes of metal products. The microscope must produce a picture that has no less than a 30-to-1 magnification.

### Purpose of the Report

The company photographer analyzed pictures taken with a 35-millimeter camera attached to Microscope A and pictures taken with a 35-millimeter camera attached to Microscope B. Tests conducted with each microscope are used as a basis for a recommended purchase.

### Definitions

Three technical terms are defined to help the reader understand the meanings intended in this report.

Depth of Field. The depth of field is the zone of sharpness in front of and behind an object being photographed. This zone can be varied by changing the lens opening on the camera.

Open Flash. Open flash is a method of controlling the amount and direction of light while a picture is being photographed. In open flash, the camera shutter is opened and a strobe light is fired one or more times from either one or several different positions. The open-flash method is used mostly to photograph stationary objects.

Grain. In photography, grain describes the appearance of the black and white specks of which a photographic print is composed. When these specks become noticeable because of excessive enlargement or overdevelopment of a negative, the picture is said to be grainy.

### Sources of Data

The cost of equipment quoted in this report was obtained from Scientific Supplies Company. Photographs used in the analysis were made by Commercial Photographers Incorporated.

### Topics for Discussion

The methods used to test the equipment, the results of the tests, and the cost of each microscope are discussed for Microscope A and then for Microscope B. The photographs (Figures 2 and 4) used in this report to illustrate the test results are arranged in pairs horizontally on the page. A stereo viewer extended to the 60-millimeter mark and rota-

**FIGURE 10-2** An example of a report for a comparative analysis

ted until the two images match up is required to make the photographs appear three dimensional.

**A 35-MILLIMETER CAMERA**
**ATTACHED TO MICROSCOPE A**

Figure 1 illustrates the camera-microscope arrangement required to photograph metal surfaces through Microscope A. In the illustration the microscope, mounted on a swing arm, is placed on a table holding the object to be photographed. A 35-millimeter camera is attached to one lens of the microscope. An electronic flash is used as a light source.

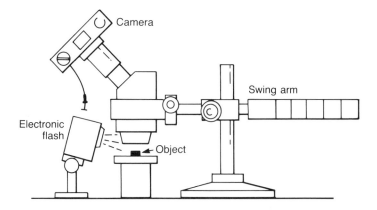

FIGURE 1. A 35-Millimeter Camera Attached to Microscope A

<u>Procedure for Testing Microscope A</u>

During the photographing process, the electronic flash was set at a low angle for strong side lighting. (See Figure 1.) To obtain sharpness in the picture, the photographer used the viewfinder located in the camera. Between exposures, the camera was manually changed from one lens of the microscope to the other lens. A high-intensity lamp was used for illumination during the adjustment required for critical focusing between pictures. The open-flash method of exposure was used to prevent vibration and to control the direction and intensity of light.

<u>Results of the Test</u>

Figure 2 shows a stereo picture photographed with Microscope A. While the pictures in Figure 2 were being photographed, three technical problems were observed: (1) Moving the camera from one lens to the other required refocusing the camera for each picture, (2) the magnifica-

**FIGURE 10-2** *(Continued)*

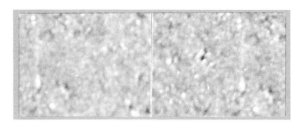

FIGURE 2.  A Stereophotograph of Metal Surfaces Through Microscope A

tion ratio of 30 to 1 was possible only through enlarging the negative,
and (3) the swing-arm attachment could not be made rigid.

The pictures in Figure 2 show that the magnification ratio has been
obtained and that the detail in the metal surface is sufficient.  The depth of
field is satisfactory.  However, the pictures are grainy and lack sharpness.

Cost of Microscope A

The initial cost of Microscope A is $600.  Each original print costs 60
cents; reprints cost 40 cents.

A 35-MILLIMETER CAMERA
ATTACHED TO MICROSCOPE B

Figure 3 illustrated the camera-microscope arrangement required to
photograph the metal surfaces with a 35-millimeter camera attached to
Microscope B.  The camera is attached directly to the microscope through
a beam-splitter phototube.  The object to be photographed is placed on
the microscope table.  An electronic flash is used as a light source.

Procedure for Testing Microscope B

During the photographing process, the electronic flash was set at a low
angle for strong side lighting.  (See Figure 3.)  Between exposures, a lev-
er was operated to mechanically change the optical beam of the microscope from one lens to the other.  The critical focusing needed between
exposures for stereo pictures was checked through a separate phototube.
The open-flash method of exposure was used.

Results of the Test

Figure 4 shows a stereo picture photographed with Microscope B.
The picture shows that the required magnification has been obtained and
that the detail in the metal surfaces is sufficient.  The depth of field is satisfactory.  The picture is sharp and shows little grain.

FIGURE 10-2 *(Continued)*

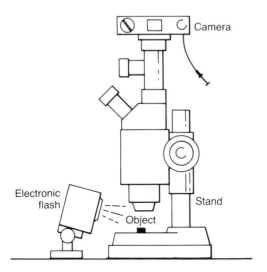

FIGURE 3.  A 35-Millimeter Camera Attached to Microscope B

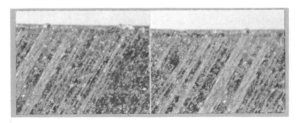

FIGURE 4.  A Stereophotograph of Metal Surfaces Through Microscope B

Cost of Microscope B

The initial cost of Microscope B is $1800.  Each original print costs 60 cents; reprints cost 40 cents each.

### SUMMARY

Stereo pictures made through two microscopes, A and B, were analyzed to determine which microscope should be purchased by Equipment Manufacturing Company.  The analysis is based upon the results of photographing metal surfaces with a 35-millimeter camera attached first to Microscope A and then to Microscope B.

FIGURE 10-2 *(Continued)*

Conclusions

The following conclusions are derived from the analysis of the two microscopes:

1. Each microscope produces pictures that have a magnification ratio of 30 to 1.

2. Each microscope produces sufficient detail of the metal surfaces and has a satisfactory depth of field.

3. Microscope A produces pictures that appear grainy and that lack sharpness.

4. Photographing objects with Microscope A is a two-step procedure in which an original negative must be enlarged to produce a 30-to-1 magnification ratio.

5. The swing-arm attachment on Microscope A does not guarantee rigidity of the equipment during the time exposure required to photograph the metal surfaces.

6. The time needed to prepare prints from photographs taken with Microscope A is 50 percent longer than the time needed to prepare prints from photographs taken with Microscope B.  The time variation results from the need to enlarge negatives in the two-step procedure previously explained.

7. The purchasing cost of Microscope A is $1200 less than the purchasing cost of Microscope B.  The cost of prints for pictures made from each microscope does not vary.

Recommendations

Although Microscope B costs $1200 more than Microscope A, Microscope B should be purchased for three reasons:

1. The two-step procedure for Microscope A increases wages paid to the photographer by 50 percent.  Therefore, in time saved, the difference in cost for Microscope B is quickly recovered.

2. The critical focusing required between stereo pictures is difficult when the equipment is not rigid.  Perfect focusing, therefore, cannot be guaranteed when Microscope A is used.

3. Pictures taken through Microscope B are sharper and less grainy; therefore,  the quality of the picture is superior.

**FIGURE 10-2**  *(Concluded)*

topics in a comparative analysis may be stated in the major topic section of the introduction if omitting these similarities makes the report incomplete. For example, if two lots being analyzed have identical measurements and almost identical contour, these similarities may not need discussion. The similarities do, however, need to be mentioned when the two lots are introduced as major topics.

5. The information in the body of the report is always logically developed, specific, and complete. Through formal classification, either written out as shown in Figure 10-2, or tabular, as in Figure 4-1, the bases used to compare each major topic are discussed. In a single analysis, where no comparison is being made, information must still be complete, accurate, specific, and logically arranged.

6. As shown in Figure 10-2, the subtopics for one major topic usually parallel those for all major topics. In other words, "Testing," "Results of Tests," and "Costs" are subtopics for both microscopes. This observation should not deter a writer from adding a necessary step to one major topic but not to another. If, for example, problems develop in only one of two methods of testing soil, these problems must be discussed.

7. The importance of graphic aids in analytical reports cannot be overestimated. Tables are used extensively. Graphs, drawings, maps, and charts also help the writer show comparisons of and contrasts in ideas for plans, problems, and equipment. The graphic aids, because they are important, must be pertinent, attractive, and simple enough to be easily interpreted. They must be numbered, titled, and integrated with the text to supplement written information.

8. An analytical report usually is long enough to justify the use of a title page and at least center and margin headings. Figure 10-2 shows the format for the first page of the report when a title page is used. The sample report also shows how margin and paragraph headings are arranged in the report.

9. When an analysis contains many formal classification tables, such as in Figure 4-1, a single table combining the information in these tables can be an effective part of the summary. The title for each column of quantitative data identifies each basis discussed. For example, if the minerals in Figure 4-1 were compared in a report according to hardness, weight, and cost, these three bases would make up the quantitative information in the summarizing table. See the subtopic "Tables" in Unit 5.

10. Conclusions and recommendations are almost always needed in an analytical report. Figure 10-2 lists conclusions and recommendations under margin headings in the summary. The information relating to one conclusion or to one reason for the recommendation is single-spaced. Double spacing is used between numbered items.

11. Conclusions are not always restatements of text information. They often are interpretations based upon the presented data. For example, in Figure 10-2, the sixth conclusion is an interpretation. In some analytical reports most or all conclusions are interpretations. Therefore, a writer uses judgment and thought combined with facts to formulate conclusions.

12. Recommendations, like conclusions, require that writers be confident of their data and of their interpretation of data. Recommendations must be unbiased. They are the results of scientific data and objective interpretations; they are not opinion.

13. Recommendations are not always conclusive. A writer, after reviewing the

contents of the report, often recommends that further studies be made or that
no action be taken. The writer usually gives reasons to justify this kind of
recommendation.

## PROOFREADING THE ANALYTICAL REPORT

The analytical report, like all reports, is checked for clarity, accuracy, completeness, and conciseness. In addition, the writer should be able to answer affirma-
tively to the following questions relating specifically to analyses:

—Are the bases upon which the analysis is being made clearly expressed?
—Are all undefined terms meaningful to the reader?
—Has every available source been researched so that information is complete?
  Have the sources of data been included in the introduction?
—Have the major topics and any necessary qualifications related to them been
  introduced to the reader?
—Is the entire report factual and scientific?
—Have conclusions been omitted from the body of the report and reserved until
  all facts are complete?
—Is the report text systematically arranged?
—Does a summary that shows how the purpose of the report was accomplished
  precede the conclusions and recommendations?
—Have graphic aids been carefully considered and used effectively where they
  are needed?
—Are informative headings included in the report?
—Are the conclusions clear, meaningful, and specific? Do they all result from a
  careful study of the report data?
—Are recommendations specific? Are they presented confidently and
  convincingly?

## SUMMARY

1. All businesses, industries, and service organizations probably use analyses more
   frequently and more extensively than any other kind of formal report. Understanding analyses and the techniques for preparing them can benefit several kinds of
   employees.

2. Analytical reports are frequently long, extending beyond a hundred pages. A group
   of analysts, for example, can be responsible for organizing data collected by
   investigators to analyze the cause or causes of a major dam break.

3. Short analyses, such as one needed after a new surge tank designed to limit waste disposal has been tested, are also required. One writer often prepares these.

4. Analyses may involve only one topic such as the dam break or the surge tank or two or more comparable topics, resulting in two general kinds of analyses: single and comparative.

5. Conclusions and recommendations are rarely, if ever, omitted from analytical reports because they are written to find a logical solution to a problem.

6. The various requirements needed to solve various problems indicate that methods for writing analyses cannot conform to rigid rules. Consequently, the writer must combine judgment and guidelines to guarantee that data is well organized, that it is valid, and that conclusions and recommendations result from objective evaluations of data.

7. Proofreading a completed report is important. It allows the writer to review data and recommendations before submitting them to the scrutiny of others.

UNIT 10

## SUGGESTED ACTIVITIES

A. Using the principles of formal classification and partition, develop two outlines, one for a comparative analysis and one for a single analysis. Select topics that are familiar to you. Before organizing the analysis, determine the purpose to be accomplished; then select the bases upon which the analysis should be made. Organize your material carefully so that you have no overlapping of information. For example, analyzing advantages of one item usually results in repeating them as disadvantages of another item. Therefore, keep your major topics and subtopics specific and clearly defined. Remember that conclusions and recommendations are essential parts of the summary.

B. Write an analytical report from one of the two outlines. If time permits, write a report from each outline.

C. Proofread the report.

D. Ask someone to read the report and to offer constructive criticism. Students often exchange papers to obtain a reader's point of view.

E. Using correct format, rewrite the report; also, prepare a title page.

F. Obtain an analytical report from the library or possibly from a local industry. Using the information taught in this unit, evaluate the report in a classroom discussion.

# The Troubleshooting Report

## OBJECTIVES

After reading this unit, you should be able to do the following tasks:

- Analyze a troubleshooting report.
- Organize a troubleshooting report.
- Write a troubleshooting report.
- Proofread a troubleshooting report.

## OVERVIEW

In a troubleshooting report, a writer explains the procedure used to analyze, describe, and correct a defect in a mechanism. The troubleshooting procedure may be so simple that it can be explained in a memorandum. Another procedure, however, such as that used in troubleshooting the *Challenger* disaster, may be so complex that many different analyses and tests must be made before the defect is located and, when possible, corrected.

The troubleshooting report is used extensively in industry to accomplish one or more of four major objectives: to help a company recognize and correct weaknesses in a new product, to prevent similar weaknesses in future products, to help technicians analyze causes of malfunctions that previously required extensive and costly analysis, and to tell a customer what caused a problem in equipment he or she has purchased.

In this unit, troubleshooting reports are analyzed, and methods for organizing and writing them are discussed. Then, a sample troubleshooting report is illustrated and explained.

# ANALYZING THE TROUBLESHOOTING REPORT

As implied in the foregoing paragraphs, troubleshooting and writing the troubleshooting report may involve one person or several. Sometimes, the technician beginning the troubleshooting process discovers that the defect must be analyzed further by a specialist in another field of technology. The following example illustrates this kind of troubleshooting problem.

**EXAMPLE**

A lawn-sprinkling system was installed for a private homeowner. After the installation, all water systems in the house and the lawn had adequate water pressure and flow. However, several years later, the rate of water flow was too low to operate the sprinklers. A lawn-sprinkler specialist was called to troubleshoot and repair the system. The specialist's analysis of the problem indicated that a blockage in the water lines preceding the stop-and-waste valve of the sprinkler system was causing reduced flow. Therefore, the city water department was called to check its water service to the house. Its technician found that the water meter was defective.

This kind of situation may result in the need for two reports or memorandums, one written by the lawn-sprinkler specialist and one by the technician from the water department.

# ORGANIZING THE TROUBLESHOOTING REPORT

As shown in the example, an analysis of a defect can require that data be obtained from one or more of the following sources:

—Manuals or instruction guides
—Manufacturers or technicians who are familiar with the mechanism
—Experts such as chemists, machinists, physicists, analysts, or engineers

Regardless of the amount of research and the complexity of procedures involved in a troubleshooting project, though, a troubleshooting report seldom includes more than the following major topics:

1. Analyzing the symptoms
2. Describing the defect
3. Correcting the defect

From these major topics and the following guidelines, the outline shown in Figure 11-1 was developed.

- Each different defect being studied requires individual interpretation.

- Troubleshooting reports are usually written for readers who are familiar with the mechanism. Therefore, definitions are seldom needed. (Exceptions must be considered. For example, the troubleshooting report about the space shuttle *Challenger* was read by many who were not familiar with the mechanism. A glossary of definitions was probably needed.)

---

**OUTLINE**

I. Introduction
   A. Background
   B. Purpose
   C. Major topics
      1. Analyzing the symptoms
      2. Describing the defect
      3. Correcting the defect

II. Analyzing the symptoms
   A. Ohmmeter check
   B. A check for power-line surge
   C. Schematic study
   D. Additional checks
      1. Possible shorts
      2. Defective capacitors
      3. Voltages

III. Describing the defect
   A. Location of the defect
   B. Causes of the defect
   C. Results of the defect

IV. Correcting the defect
   A. Installation of new parts
      1. Picture tube
      2. Tube socket
   B. Additional procedures
      1. Isolate the picture-tube heater
      2. Install a separate transformer
      3. Ground the spark-gap ring

V. Summary

---

**FIGURE 11-1**  An example of an outline for one kind of troubleshooting report

- Only one individual analyzed the defect. Thus, a "Sources of Data" section was not needed. (It is needed if other personnel — for example, those previously mentioned — participate in or contribute to the investigation or the report.)

- Subtopics to the three major topics previously listed show the steps involved in developing the major topics. (Several subtopics, probably including second- and perhaps third-level subtopics, are needed when an investigation is complex.)

- The "Summary," a recapitulation of data, completes the outline. (Other outlines may include "Conclusions" and "Recommendations." See the next section.)

# WRITING THE TROUBLESHOOTING REPORT

The report in Figure 11–2 was developed from the outline in Figure 11–1. It shows that both analysis and steps in a process are the bases for the text in the troubleshooting report. The following comments explain how the report in Figure 11–2 was written and offer suggestions for writing other troubleshooting reports.

1. Judgment (Unit 2) is more valuable than rigid guidelines as an aid in writing effective troubleshooting reports.
2. The report in Figure 11–2 is only an illustration of the three major topics to be considered for all troubleshooting reports. It is not a pattern to be consistently adopted.
3. "Analyzing the Symptoms" includes all data needed for a detailed explanation of the procedures involved in the analysis. Each procedure becomes a subtopic. Therefore, a report concerning a complex malfunction can have many first-, second-, and possibly third-level subtopics. These subtopics and the amount of text written about each determine the number and kinds of headings used. See "Placing Headings in Reports" in Unit 7.
4. "Describing the Defect" includes all the information a reader needs to understand the defect. It usually describes the precise location, often emphasized by the use of graphic aids; the causes of the defect; and the conditions resulting from the defect.
5. In "Correcting the Defect," each procedure used to correct the defect is discussed in detail. Sometimes, attempts at making corrections fail. If so, details of the attempts are explained. Then, the resolution — such as replacing the part or sending the mechanism to another expert or its manufacturer for further study — is stated. Occasionally, as in the example of the lawn sprinkler, a person or company not originally involved may be

*Title could
be placed on
a title page*

**TROUBLESHOOTING THE CAUSE
FOR A BURNED-OUT PICTURE-TUBE HEATER
ON A BLACK-AND-WHITE TV SET**

*Background*

On October 22, 1985, a customer brought a black-and-white television set to Bill's TV Repair and said that the light on the screen had faded away after months of trouble-free service. Andrew Garry, the technician assigned to repair the set, installed and burned out a new picture tube before he discovered and corrected the defect.

*Purpose*

*Major topics*

The following detailed explanation of the method used to locate and repair the defect will help other TV technicians check for similar defects before they replace the picture tube. Garry analyzed the symptoms. He then followed a logical step-by-step procedure to locate and correct the defect.

**ANALYZING THE SYMPTOMS**

The fading of light on a television screen indicates that the picture-tube heater has probably burned out. Garry verified this analysis by checking the heater connections with an ohmmeter.

A picture-tube heater usually burns out for one of two reasons: (1) a power-line surge causes a momentary overload or (2) the picture tube was defective when the set was built, but the defect was not obvious before the tube was installed.

A Check for Effects of Power-Line Surge

A power-line surge that burns out a picture-tube heater may also burn out one or more of the smaller tubes in the set. Therefore, Garry, after removing only the picture tube, turned on the set. Every remaining tube was functioning properly.

Schematic Study

*Topic
sentence*

After the small-tube check showed the heaters were good, the schematic was studied. All tubes in the set were

**FIGURE 11–2** An example of a troubleshooting report

wired in parallel, and one end of the heater was tied to the chassis ground. After these checks also showed no defects in the smaller tubes, a new picture tube was installed. The set worked perfectly. It was operated at Bill's TV Repair continuously for 8 hours a day for three days. On the fourth day, however, the newly installed picture tube burned out.

Additional Checks

*Topic sentence*

Assuming that two consecutive picture tubes would not be defective, Garry made a thorough check of all components. Using the schematic as a guide, he checked the entire chassis for shorts or near shorts. Capacitors were disconnected and checked. Voltages were measured. No defects were found.

After all previous checks failed to locate the defect, the picture-tube socket, which is rarely defective, was examined. All the clips in the socket that contact the picture-tube pins were carefully inspected. At this point, the defect was found in the socket.

### DESCRIBING THE SOCKET

Within the socket, a metal strip connects the heater wire through the tube socket to the heater pin. This strip had fatigued and burned, creating an open circuit.

Causes of the Defect

The picture-tube socket, which is preassembled at the factory, contained built-in protective arc gaps. These gaps had triangular-shaped points that formed arc gaps with several of the contact clips. The ground contact of the arc gap was connected to chassis ground by means of a common brass ring.

Results of the Defect

High-voltage transients had now found it easier to arc inside the tube to the grounded heater than outside the tube to the grounded ring. Therefore, the spark gap protecting the cathode was burned off. The mechanical shock to the heater wire caused by repeated arcing finally burned the heater wire in two. A check of the second burned-out pic-

**FIGURE 11-2** (*Continued*)

ture tube showed that the picture-tube heater had separated because of a high-voltage arc.

### CORRECTING THE DEFECT

To repair the defect, Garry installed a second new picture tube and a new tube socket containing a spark-gap ring. He also used additional procedures to prevent a recurrence of burnouts. He isolated the picture-tube heater from all other tubes in the set and installed a separate transformer to heat only the picture tube. In addition, he grounded the spark-gap ring directly to chassis ground. Thus the high-voltage transients had no need to arc to the heater at any time.

### SUMMARY

*Recapitulation*

This report explains in detail the procedure used at Bill's TV Repair to analyze, locate, and correct the cause of a picture-tube-heater burnout in a black-and-white TV set. The analysis was difficult and expensive because the defect was located in the picture-tube socket, a component that is rarely defective.

**FIGURE 11-2** *(Concluded)*

required to correct the defect. If so, a "Correcting the Defect" section is probably not needed in the report because a concluding statement in "Describing the Defect" tells the reader what action was taken.

6. Only judgment can help a writer determine the need for graphic aids. None are needed in the report in Figure 11-2 because it is written for technicians who are familiar with black-and-white TV sets.

7. The writer must decide whether any components related to the defect are unfamiliar to a potential reader. If so, a physical description of them is needed. It may be the first major topic of the report or incorporated within the text where it is first mentioned. (A graphic aid may simplify the description.)

8. Specific, detailed explanations of work performed to analyze, describe, and correct the defect are essential. Assumptions and opinions are never substitutes for careful analysis, research, and investigation. Therefore, a writer avoids using them unless they are based upon logical analysis and can be justified.

9. In addition to the recapitulation given for the report in Figure 11–2, conclusions and recommendations (Unit 10) are sometimes required. They are used primarily when information is inconclusive or when additional research is needed to correct the defect. For example, a mechanism that has proved satisfactory in the laboratory shows defects in practical applications. The defects may be recognized when the mechanism is returned to the laboratory. However, methods for correcting them require experimentation beyond that previously performed. The technician who field-tested the mechanism may summarize his findings and recommend further experimentation.

## PROOFREADING THE TROUBLESHOOTING REPORT

The troubleshooting report is proofread according to the rules listed for all reports in Unit 8. In addition, because the reader usually understands the mechanism being discussed but not the cause of the malfunction, the following questions are considered:

—Is the purpose stated clearly and specifically?
—Have outside sources of information, if used, been listed in the introduction?
—Have the major topics been listed and, if necessary, qualified in the introduction before they are discussed in the text?
—Has the writer carefully analyzed the reader's understanding of the mechanism and the problem so that all pertinent information is included but unnecessary information is omitted?
—Is the procedure used to analyze the problem and to locate and correct the defect detailed and specific?
—Is the report developed systematically so that the reader understands the logic that the technician followed to locate and correct the defect?
—Is all information presented scientifically and objectively?
—Has the text been summarized?
—Have conclusions and recommendations been considered and included if they are needed?

## SUMMARY

1. A troubleshooting report explains the procedure used to find and repair a defect in equipment.

2. A simple procedure and repair, especially one for only in-house readers, may be written as a memorandum. Lengthy, complex procedures require a formal report.

3. Three major topics — analyzing the symptoms, describing the defect, and correcting the defect — are components of all troubleshooting reports.

4. Troubleshooting reports may result from one person's investigation and repair. In contrast, the problem may be complex enough that two or more experts from more than one discipline become involved.

5. Graphic aids often give valuable support to text in troubleshooting reports. They must be carefully coordinated with the text.

6. Conclusions and recommendations are part of the troubleshooting report when they can help prevent future problems or when the current problem requires further analysis.

7. Proofreading for logical development, adequate detail, clarity, completeness, and accuracy is important.

UNIT

# SUGGESTED ACTIVITIES

A. Take notes as you follow through a troubleshooting problem in the laboratory or on the job. Especially note specific thought processes and investigations used to analyze the problem and the details of procedures used to locate and correct a defect. If any wrong analyses were made or any processes that proved unsuccessful were followed, make a note of them and include them in the report to be written later. (An explanation of unsuccessful procedures helps other technicians avoid them and thus saves time and money.)

B. After carefully analyzing the purpose to be accomplished and the reader's background, prepare an outline for a troubleshooting report.

C. Write a report from the foregoing outline.

D. Proofread the report; then ask a person (preferably a classmate, a laboratory assistant, or an instructor) to offer constructive criticism.

E. Check headings and graphic aids to determine whether they are meaningful. Using correct format, prepare the final draft of the report.

# The Examination Report

## OBJECTIVES

After reading this unit, you should be able to do the following tasks:

- Analyze an examination report.
- Organize an examination report.
- Write an examination report.
- Proofread an examination report.

## OVERVIEW

Examination reports, important in nearly all industries, are primarily recorded data obtained from an examination of mechanisms or conditions. A technician on an observation field trip often reports the data in a memorandum (Unit 13) and mails it to his or her supervisor. In contrast, an examination of a complex mechanism or set of conditions may require a detailed, formal report.

In this unit, examination reports are analyzed; then techniques for organizing and writing them are explained and illustrated.

# ANALYZING EXAMINATION REPORTS

Most routine examination reports are written by one person. However, the mechanism or set of conditions may be examined by a team of experts, all of whom then contribute to the report. The following examples explain four circumstances under which examination reports may be required.

1. A company complains about the performance of equipment purchased from a manufacturer. A technician or engineer representing the manufacturer examines the equipment under field conditions to determine whether the complaint is justified.
2. A company must purchase equipment to perform a particular type of work. An employee of the company may be sent to various manufacturers to examine equipment and report his or her observations. For example, a company needs a motor to rotate an antenna at a speed of 900 revolutions a minute within a specified voltage and temperature range. A technician or engineer may watch many operating motors, using required specifications as a base, and record information about each motor.
3. A company has developed a new machine to replace an old one. The old and new machines may be operated at the same time and observed so that the company can compare the efficiency of each.
4. The condition of a forest, a road, a building, a mine entrance, a fire-protection system, and many similar conditions are frequently recorded in an examination report.

# ORGANIZING THE EXAMINATION REPORT

As the foregoing examples imply, examination reports differ from one another in subject matter and length. Some are similar to analytical reports but are less complicated because the information is obtained only from personal observation. When only one mechanism or one condition is examined, the information is often written as a memorandum. A complex examination involving a comparison, a set of specifications, or an examination of more than one item is written as a formal report. A sample outline for an examination report concerning the performance of two hard-rock drill bits is illustrated in Figure 12–1.

The following guidelines explain the logic used in the development of the outline.

- Examination reports are usually written for department supervisors who are familiar with the equipment or conditions being examined. Therefore, definitions are rarely needed.

I. Introduction
  A. Background
  B. Purpose
  C. Major topics
    1. Drill bit No. 1345
    2. Drill bit No. 6894

II. Examination of drill bit No. 1345
  A. Operating conditions
    1. Location
    2. Rock formation
    3. Operation equipment
  B. Observations
    1. Penetration rate
    2. Wear
      a. Profile
      b. Outer diameter
      c. Inner diameter
    3. Salvage

III. Examination of drill bit No. 6894
  A. Operating conditions
    1. Location
    2. Rock formation
    3. Operation equipment
  B. Observations
    1. Penetration rate
    2. Wear
      a. Profile
      b. Outer diameter
      c. Inner diameter
    3. Salvage

IV. Comments from foremen
  A. Penetration
  B. Bit life

V. Summary

**FIGURE 12-1**  An example of an outline for an examination report

- The text is based upon personal observation. Therefore, a "Sources of Data" or "Omitted Data" section is not part of the report introduction.

- The equipment or conditions being examined are the major topics in Figure 12-1 and in all examination reports that do not include specifications. When equipment is being examined to determine whether it meets certain specifications, "Specifications" is probably the first major topic. Each mechanism examined follows as a major topic.

- A summary is given in all examination reports except memorandums.

- Conclusions and recommendations are not included in examination reports. Once they are added, the report becomes an analysis.

# WRITING THE EXAMINATION REPORT

An examination report must be complete, correct, and specific. Therefore, written notes should be made while the mechanism or condition is being examined. A list of items to be observed, written before the examination is made, makes note taking easier.

Figure 12-2 illustrates the report written from the outline in Figure 12-1. The outline and report were prepared by a technician from notes taken during a field trip to a mine. The following comments explain how the report was developed. They also justify the use of a table in the summary.

---

### INTRODUCTION

From June 14 to 16, 1979, a technician from Morris Drill Bit Manufacturers examined two newly designed hard-rock drill bits operating at Monarch Mines. This report explains the conditions under which drill bits No. 1345 and No. 6894 were operating and the observations made by the technician. Comments of mine foremen who operated the bits are also summarized.

### EXAMINATION OF DRILL BIT NO. 1345

On June 14, 1979, drill bit No. 1345 was operated continuously for 8 hours.

Operating Conditions

The bit was operating 12 feet underground in drill hole No. 85. The rock formation being drilled was hard, dense quartz. The drill hole was equipped with a 34-core drill and a 20-gallon pump. The core drill and pump were powered by a 4-cylinder air motor.

---

FIGURE 12-2   An example of an examination report

Observations

The bit drilled a total of 30 feet in 8 hours. The average penetration was 3.75 feet an hour. In the first 8 feet of drilling the bit showed extreme wear on the protruding profile. For the remaining 22 feet drilled, the bit showed fast, continuous wear. After 8 hours, the bit, which was nearing the end of bit life, was retired.

The outer diameter of the bit looked good but was 0.005 inch under size. The inner diameter showed no wear. Throughout the 8 hours, the bit performed without vibration. Core recovery was 100 percent.

## EXAMINATION OF DRILL BIT NO. 6894

Drill bit No. 6894 was observed for 8 hours on June 15 and for 2 hours on June 16, 1979.

Operating Conditions

The bit was tested in hole No. 86 at a starting depth of 11 feet. The material being drilled for the first 10 feet was a soft, abrasive schist. The remaining 62 feet drilled consisted of a harder, solid schist containing intermittent quartz seams. Hole 86 was equipped with a No. 2 air-core drill and a 20-gallon pump. The core drill and pump were powered by a 4-cylinder air motor.

Observations

Drill bit No. 6894 drilled 56 feet in 8 hours on June 15. On June 16, it drilled an additional 16 feet in 2 hours before it was retired. The average penetration rate was 7.2 feet an hour. After the first 10 feet of drilling, the protruding profile of the bit showed some wear. After 32 feet of drilling, the profile was flat. However, the bit drilled an additional 40 feet. During this last drilling, the profile showed slow, even wear until, when the bit was retired after 10 hours, the profile was worn away. The outer diameter had worn only 0.005 inch. The inner diameter showed no wear. The bit drilled without vibration until it was retired. Core recovery was 100 percent.

## COMMENTS FROM FOREMEN

Two foremen who operated the drill bits at the mine, Mr. Carl Heisner and Mr. William Froman, expressed their opinions concerning the two drill bits. A summary of these comments follows.

1. The penetration rate of both bits was excellent.

**FIGURE 12-2** *(Continued)*

2. The life of the drill bits was shorter than the foremen had anticipated.
3. Reduction of the penetration rate may have added more footage before bit life ended.

### SUMMARY

An examination of two newly developed bits for hard-rock drilling was made at Monarch Mines from June 14 to June 16, 1979. The operating conditions and the performance of drill bits No. 1345 and No. 6894, explained in detail in the report, are summarized in Table I. Core recovery for each bit was 100 percent.

### TABLE I

### BIT LIFE AND PERFORMANCE
### FOR DRILL BITS NO. 1345 AND NO. 6894

| Bit No. | Type of rock | Length of life | Total feet drilled | Average feet an hour |
|---|---|---|---|---|
| 1345 | Hard, dense quartz | 8 hours | 30 | 3.75 |
| 6894 | Soft to hard schist | 10 hours | 72 | 7.20 |

FIGURE 12-2 *(Concluded)*

1. The report is long enough to justify a title page. Therefore, the first page of the text begins with the introduction.
2. The information needed to clarify the background, purpose, and major topics is brief and interrelated. For this reason, only one short paragraph is used in the introduction.
3. The examination of each drill bit is discussed as a separate major topic. This organization is similar to that used in comparative analyses. Two subtopics,

one explaining the conditions under which the bits were operated and one telling what was observed, are used for each major topic.

4. If specifications were part of this report, subtopic arrangement would be planned so that performance would be related to specifications. For example, if rate of engine speed, power, and torque were included in specifications for a motor, these terms and an explanation of testing conditions would be probable subtopics in the examination report.

5. Only rarely is a graphic aid introduced in a summary. However, Table I does not relate to any one major topic; it summarizes the important data obtained from the examination. This unconventional use of a table illustrates the importance of incorporating graphic aids where they are needed.

6. The arrangement shown in Table I may be reversed. The bits being examined could be placed horizontally and the items observed arranged in the vertical left-hand column. This latter arrangement may be preferable in a comparative report or one that includes specifications.

7. In Figure 12-2, the performances of the drill bits were not being compared. The bits were examined as separate mechanisms so that the manufacturer could determine the efficiency of each newly developed drill bit.

In a report that includes specifications, conclusions would be made to show how the examined equipment did or did not meet the required specifications. In addition, a recommendation to purchase a piece of equipment or to make further studies would probably be expected.

# PROOFREADING THE EXAMINATION REPORT

The examination report, like any document, is read at least three times before the final draft is prepared. In the first reading, the writer determines whether the text is logically developed, has continuity, is meaningful, and can be read easily. The writer also determines whether the purpose is clearly expressed and has been accomplished.

In the second reading, the writer checks for errors in grammar, sentence structure, spelling, punctuation, capitalization, and the use of numbers. The following questions are also considered.

—Have all necessary data been included?
—Are essential facts and figures specific and accurate?
—Have necessary graphic aids, especially tables, been included?
—Are graphic aids simple and neat? Are all parts clearly labeled? Is the title informative?

—Are all figures in the tables typed or written clearly so that they cannot be misread?

In the final reading, the writer makes sure that any revisions made while the report was proofread have not changed the logical development, continuity, or clarity.

# SUMMARY

1. An examination report is a logically organized record of on-the-job observations. Employees investigating major or minor accidents, forest infestations, or natural disasters report only what they observe; they prepare examination reports. Analysts, either the original observer or someone else, use these observations as a basis for analytical reports.

2. Because examination reports tell only what is happening, they are usually easy to write, especially for an observer who keeps step-by-step notes while he or she is observing.

3. Examination reports are nearly always prepared for people already knowledgeable about the subject, not for general readers.

4. Examination reports do not contain conclusions and recommendations. They contain only observed facts, not interpretations of facts.

5. Because the facts from examination reports are often used for later analyses, the data must be well organized, clearly presented, and complete. Therefore, a person writing these reports is obligated to concentrate upon these requirements while proofreading his or her report.

UNIT **12**

# SUGGESTED ACTIVITIES

A. If you are studying subjects related to mechanisms, plan to observe two or more operating pieces of equipment. Use one or more specifications as a basis for the observation. Plan only to observe the equipment, not to make comparisons.

   If you are studying nursing skills, fire prevention, food service, carpentry, or similar subjects, plan an examination of one or more conditions. For example, make a sterile-technique inspection, a cleanliness inspection, a sprinkler-system inspection, an inspection of a building being constructed, or some similar inspections. Have a set of specifications to be used as a standard for the inspection.

B. Before examining the mechanisms or conditions, prepare a list of items to be observed including, if necessary, the location, working conditions, and other factors that may affect the performance of a mechanism or the results of an inspection.

C. While examining the mechanisms or conditions, write complete, specific notes of everything observed.

D. Prepare a title and an outline for the planned report.

E. Using the outline, prepare the report. Do not write this report as a memorandum.

F. Proofread the report according to the procedure recommended in this unit; then ask someone unfamiliar with the information to offer constructive criticism.

G. Determine whether a title page is needed, and prepare the final draft of the report.

H. Exchange reports within the classroom and make constructive written or oral criticism.

# Interoffice Communications

## OBJECTIVES

After reading this unit, you should be able to do the following tasks:

- Prepare and write memorandums.
- Prepare and write form reports.
- Describe the uses and types of memorandums and form reports.
- Prepare and write letter reports.

## OVERVIEW

The reports discussed as formal reports in Units 8 to 12 are not written at regular intervals but only when a need for them exists. They are prepared for personnel in a local office, for affiliates, or for other interested readers. In contrast, one- or two-page daily, weekly, or routine communications to be read only by personnel in a local office are frequently written as memorandums, memorandum reports, or form reports. Other short reports, primarily summarizations of lengthy analyses, are prepared for other companies or for executives within the company. These reports are more formal and are presented in letter format.

In this unit, the following interoffice communications — memorandums, memorandum reports, form reports, and letter reports — are illustrated and explained.

# MEMORANDUMS

The term *memorandum*, sometimes referred to as *memo*, has a variety of uses and formats in modern industry. A memo is a fairly informal communication, usually written for interoffice circulation. It is normally used to send an informal message to another employee or to submit a short, informal report.

## Preparing Informal Messages

In some companies, memorandum forms are quarter or half sheets of paper labeled as memos. These forms are used for short messages that have only temporary significance. See Figure 13-1. In some companies, the memorandum form is approximately two-thirds the size of a standard sheet of typing paper and has two divisions: a section for an original message and a section for a reply, as in Figure 13-2. In other companies, a memorandum form is an 8 1/2-inch by 11-inch sheet of paper. Printed near the top of the page are the company's name and, sometimes, the word *memorandum*. Four additional words — *To, From, Subject*, and *Date* — are also printed.

Few, if any, technical writers have difficulty preparing informal messages similar to those shown in Figures 13-1 and 13-2 or in understanding the uses for

```
              INTERNAL
              MEMORANDUM


     Date:      July 16, 1987
       To:      All Supervisors

     From:      James Owens, Inventory Control Supervisor
  Subject:      Materials Transfer Procedures

      May we ask you for your cooperation?  Machinery and
  supplies are being moved from one department to another
  without proper authorization. As a result, Inventory Con-
  trol is having difficulty keeping accurate records.

      Please ask your employees to follow the approved pro-
  cedure given in the Inventory Control Procedures Manual for
  transferring equipment.
```

FIGURE 13-1  An example of a short-form memorandum

*RAPIDFORMS* NO.11003   REORDER FROM REGENT STANDARD FORMS, INC., INTERSTATE INDUSTRIAL PARK, BELLMAWR, NJ 08031

**LETTER-LIMINATOR**   Sender: snap out yellow copy only. Send white and pink copies with carbon intact.

TO   **George Nichols**
     **Circle Thread Company**
     **17 Nelson Avenue**
     **Lakewood, NJ 12300**

FROM
     **Mark Hilton**
     **Ramson Machine Company, Inc.**
     **1250 Broadway**
     **Milwaukee, WI 00321**

SUBJECT:   Specifications for Model K-12

Fold here

DATE **5/29/86**                    **MESSAGE**

We received your specifications for Cutter K-12.
However, the quoted price does not indicate whether you
are ordering from Lot 99 or Lot 103.

SIGNED *Mark Hilton*

DATE **6/5/86**                    **REPLY**

Mr. Hilton:
     Our engineers recommend your specifications for Lot
number 103.

SIGNED *George L. Nichols*

RECIPIENT: RETAIN WHITE COPY, RETURN PINK COPY

**LETTER-LIMINATOR**

FORM 11003 REGENT FORMS, BELLMAWR, N.J.08031
SENDER: SNAP OUT YELLOW COPY ONLY. SEND WHITE AND PINK COPIES WITH CARBON INTACT.

**FIGURE 13-2** An example of a memorandum that contains space for a reply

the short, self-explanatory format. For this reason, only one general guideline for preparing informal messages is given: Telescopic writing and incomplete sentences are avoided.

The 8 1/2-inch by 11-inch format shown in Figure 13–3 may also be used for informal messages, especially lengthy ones such as a list of instructions; company policies, rules, or objectives; or comments related to job performance. It is also used for memorandum reports.

## Preparing Memorandum Reports

Memorandum reports, mentioned previously in Units 11 and 12, are written when required information about a process, an analysis, an examination, or a mechanical defect is not sufficient for a formal report. See the brief examination report in Figure 13–3. Regardless of the length or content, memorandum reports are informal messages. For this reason, they are sent only to personnel, usually managers or supervisors, within the organization.

Because a standardized memorandum page has a simple format and does not include an introduction or a summary, people sometimes write about their subject carelessly. However, all kinds of reports are more effective when carefully prepared. Therefore, the following guidelines for memorandum reports should be closely followed.

- Standard English is used in all memorandum reports. Informal English is reserved for memorandum messages, instructions, or notes. Telescopic writing (omissions of *the*, *a*, *an*), abbreviations, and symbols are avoided.

- Technicians submit short, daily reports as memorandums instead of as form reports (see Figure 13–5 appearing later in this unit) or as letter reports (see Figure 13–8 later in this unit) only when company officials approve.

- The text is organized so that the information is presented logically, completely, and clearly. All statements are written as complete sentences.

- The text is written specifically and objectively. Personal pronouns *I*, *we*, and *you* are avoided unless the memorandum author is certain they are acceptable to supervisory personnel.

- Typewritten reports are always preferable to handwritten ones. However, if handwritten memorandum reports are acceptable and necessary, the handwriting must be legible. Handwritten numbers must be so distinct that they cannot be misinterpreted by the reader. Spacing between lines and words must be arranged so that the format is neat and attractive.

- If double spacing is not used between handwritten paragraphs, all paragraphs must be indented five letter spaces.

- Double spacing must be used between margin headings and text.

---

COMPANY NAME

MEMORANDUM

To:          Taylor Leaming
             Plant Engineer

From:        Joseph Romano

Subject:     USE OF THE FRONT-END LOADER AT PLANT A

Date:        January 3, 1986

As you requested, I am submitting this memorandum to show
seven uses for the front-end loader and the percent of time
the loader was operated at Plant A during 1985.

Uses of the Front-End Loader

During 1985, the front-end loader was used to perform seven
different kinds of work:

1.   To move raw materials
2.   To scrape and clean roadways
3.   To remove winter snow from parking lots and roads
4.   To unload delivered merchandise from trucks and rail-
     road cars
5.   To push railroad cars into position on spurs
6.   To load scrap metal and waste into trucks
7.   To lift heavy items

Percent of Time Loader Is Used

During 1985, the front-end loader at Plant A was operated
90 percent of each 8-hour shift during the 9-month season.
It was operated 100 percent of each 8-hour shift during
the 3-month road-repair season.  While the loader was operating 100 percent of the time, many materials normally
moved by the front-end loader were moved manually. Two or
three workers were needed to move heavy items.

CC:  Alice Smythe and Tom Arden

---

**FIGURE 13-3**  An example of an examination report prepared as a memorandum report

# Writing a Memorandum Report

Figure 13–3 illustrates a simple examination report written as a memorandum. The following comments relate to generally acceptable procedures for writing memorandum reports such as the one in Figure 13–3.

1. The name of the person to whom the report is directed, the writer, and the subject will be equally indented from the left-hand margin. This indentation is usually 10 to 12 spaces.

2. The name of the person to whom the report is directed is placed after *To*. If space permits, the department or title may be placed on the next line, below the person's name. A double-space precedes the next item.

3. The name of the sender follows the word *From*. *Subject* is printed two spaces below *From*.

4. The subject of a memorandum report is similar to the title of a formal report but is condensed so that it preferably does not extend beyond the subject line shown in Figure 13–3. All letters in the subject are capitalized. A double space separates *Subject* from *Date*.

5. The text always begins two spaces below the date line.

6. A short introductory statement similar to the one shown in the figure precedes the text. This introductory statement, which can be varied, is used to introduce the content of the report. Formal introductions are not required in memorandum reports.

7. The text should be logically organized. Each major division should be preceded by a margin heading. Occasionally, paragraph headings are also included. Double spacing above and below the margin headings separates the headings from the text.

8. Memorandum reports are single-spaced; however, except in handwritten text, double spacing is used between paragraphs. The first line of each paragraph may be indented five spaces or may be flush with the left-hand margin.

9. Usually the original of a memorandum is sent to only one person. Others who receive copies are listed at the end of the report.

10. Unless company policy requires it, the author does not sign a memorandum report. The author's name has already been entered at the top of the page. Some writers do prefer to initial the memo, however, immediately following the typed name.

11. Graphic aids are rarely used in memorandum reports. If they are considered necessary, a short table or figure may be included.

12. A summary is omitted from memorandum reports.

13. A recommendation is occasionally included. However, when a writer must justify recommendations, he or she writes a formal report or a letter report.

# FORM REPORTS

In place of the general memorandum, which can be used for a variety of subjects, many organizations prefer specific forms for specific kinds of informal reporting. For example, medical and other scientific laboratories usually have several different forms on which test procedures, observations, or data concerning experiments are recorded. Engineering, manufacturing, or construction firms that have several employees performing various activities away from the home office rely upon data submitted on preprinted forms for periodic information concerning these activities. Two preprinted forms from one construction company are shown in Figures 13-4 and 13-5.

Numerous other kinds of form report formats exist. However, all of them have either open space for text, as shown in Figure 13-4, or preprinted headings requiring short answers, as shown in Figure 13-5.

## Preparing Open-Text Form Reports

A form title and possibly a form number designating the kind of information to be recorded are preprinted at the top of open-text report forms. Below the title, several headings followed by blank lines are usually preprinted to help the reporter record routine information easily and completely. The rest of the page, like pages in most memorandums, is open or filled with blank lines. The reporter must prepare the text for this section of the report. Six guidelines for effectively preparing the text are given next.

- Often, informal language is used in conversations or in brief, informal notes within an organization. However, standard English in all reports is recommended. The use of standard English implies that the author has respect for the reader and has self-confidence.

- The text is written specifically and objectively. Telescopic writing, abbreviations, and symbols are avoided. Only complete sentences are used.

- The text is organized so that information is presented logically, clearly, and completely.

- Usually, only one subject is discussed. Therefore, headings are not likely to be needed. However, if they seem to contribute to understanding, they are included.

- Many form reports, especially field reports, must be handwritten. The handwriting must be legible. If necessary to ensure readability, the text is

NATIONAL MECHANICAL CO., INC.

DAILY FIELD REPORTING FORM NO. 101176-A

DATE:___  ___  ___     JOB NO:_____   PROJECT:_____

WEATHER:_____   A.M. _____   LOCATION:_____

                       P.M._____   CRAFT:_____

PRESENT AT SITE:_____

_____

_____

THE FOLLOWING WAS NOTED:_____

_____

_____

_____

_____

_____

_____

_____

_____

_____

_____

_____

_____

_____

_____

_____

_____

_____

_____

_____

DISTRIBUTION: Retain copy for field files - Mail (1) one

copy to Home Office (send Original)

SIGNATURE:_____

**FIGURE 13-4** An example of an open-text report

| NATIONAL MECHANICAL CO., INC. | DAILY EXTRA WORK SUMMARY AND RENTAL RECAP BILLING FORM |
|---|---|

PROJECT_____ DATE_____

DESCRIPTION OF WORK_____

_____

_____

FOR_____AUTHORIZED BY_____

| LABOR | | STRAIGHT TIME | | OVERTIME | | AMOUNT |
|---|---|---|---|---|---|---|
| BADGE NO. | CLASSIFICATION | HOURS | RATE | HOURS | RATE | |
| | | | | | | |
| | | | | | | |
| | | | | | | |
| | | | | | | |
| | | | | | | |
| | | | | | | |
| | | | | | | |
| | | | | | | |

| EQUIPMENT USED | | | HOURS | RENTAL RATE | AMOUNT |
|---|---|---|---|---|---|
| CODE NO. | KIND | ATTACHMENTS | | | |
| | | | | | |
| | | | | | |
| | | | | | |
| | | | | | |
| | | | | | |
| | | | | | |

| SUPPLIES AND MATERIALS | | | UNIT PRICE | AMOUNT |
|---|---|---|---|---|
| QUANTITY | UNIT | DESCRIPTION | | |
| | | | | |
| | | | | |
| | | | | |
| | | | | |

| | SUMMARY | AMOUNT |
|---|---|---|
| SIGNED_____ | LABOR_____ | |
| BY_____ | EQUIPMENT RENTALS | |
| | SUPPLIES AND MATERIALS | |
| APPROVED_____ | PAYROLL TAXES AND INSURANCE @ | |
| COMPANY | | |
| BY_____ | OVERHEAD AND MARGIN @ | |
| | TOTAL | |

**FIGURE 13-5** An example of a short-answer report (Source: National Mechanical Co., Inc., Salt Lake City, Utah)

printed. Adequate space separates one word from another. Rules for correct punctuation, spelling, and paragraphing are observed. Numbers written as figures must be so distinct that they cannot be misinterpreted.

- If lines like those for the text shown in Figure 13–4 are omitted, the reporter is obligated to keep all lines of text evenly and clearly spaced.

## Preparing Short-Answer Form Reports

A few short-answer form reports may require no more than check marks or a *yes* or *no* in the blank following each preprinted heading. In the preparation of short-answer reports, the following guidelines apply:

- Even though briefly stated, information must be accurate, easily understood, concise, and complete.

- Words and figures are written so clearly that they cannot be misread. Words are often printed to ensure readability.

- Abbreviations and symbols are used only if they are approved by supervisors in individual organizations. Even then, a technician does not create original abbreviations but uses only those in general use within the organization.

- A technician should learn whether an employer wants all blanks filled even though some answers cannot be given. If so, *no, none,* or a similar word is written;  an omission can suggest the topic has been overlooked.

# USES FOR FORM REPORTS

Here are the types of reports most frequently written as form reports:

1. Simple or routine examinations of equipment or conditions
2. Routine laboratory tests that do not require detailed analysis
3. Progress reports
4. Periodic reports
5. Justification reports

Any of these reports may also be written as a more formal letter report explained later in this unit.

## Examination Reports

Examination reports were discussed in Unit 12 as formal reports. However, a simple examination of one mechanism or one condition may be written as a memorandum report or a form report, depending on the preprinted format used.

## Laboratory Reports

The procedures and results of an occasional laboratory test are often submitted on a memorandum or a report form. The information in the report is usually organized into three parts.

1. A description of the materials used in the test
2. Test procedures
3. Test results

## Progress Reports

In most industries, progress being made on continuing projects such as constructions, installations, repairs, and sample testing must be reported at regular intervals, usually every one or two weeks. Technicians who perform only one type of work on a project can usually report their progress to a supervisor in a memorandum or a form report. The supervisor, who reports the progress on the entire project, often prepares a longer, formal report.

The informal progress report, shown in Figure 13–6, may be divided into two or three parts. The first part summarizes the work completed at the end of the previous reporting period. The second part gives a detailed explanation of the work completed since the last reporting period. If anticipated progress has not been made between reporting periods, reasons for incomplete work are also included here.

An outline for future plans is sometimes requested or voluntarily submitted. This outline constitutes the third part of the report.

## Periodic Reports

A periodic report may be identical to a progress report because it explains work done during a specified period of time. However, a periodic report may be a description of various activities, often those performed and reported daily. For example, a public health nurse assigned to home visitations may be asked to record, each day, information concerning each patient he or she visited. The organization of the subject matter in the report is determined by its content. When many different jobs are involved, each job may be part of a margin heading. Then

**CP&H**   CENTRAL PLUMBING AND HEATING COMPANY

Date:      March 24, 1986
To:      Jerry Thompson, Projects Superintendent
From:      Mel Harris, Journeyman Plumber
Subject:      Progress Report for March 17 to March 21

Work to March 14 Summarized

Last week's progress report showed that plumbing work at
the Lakeview Hospital had progressed to the point that no
more can be done until painting is completed.  The heating
system cannot be installed until duct work is completed and
coils are installed.

Progress from March 17 to March 21

I checked Monday, March 17, and found that work delays
identified above still prevent further installation at
Lakeview Hospital.  (The heating coil had been installed on
Monday but must be changed.  Its position does not allow
space for our work to be done.  I have submitted a request
for this change.)

I began work at the rhinoceros and elephant house at Wil-
derness Zoo.  I ordered the pipe needed and started installation of the major drains, connecting 6 of the 8.  I discussed the original layout of the watering trough with the
zoo curator, suggesting it seems too narrow to accommodate
the tusks on a rhinoceros.  The curator has now authorized
change to a wider trough.

On Friday, March 21, I again checked the progress of other
work at the hospital and learned that plumbing still cannot
be completed.

Projection for March 24 to 28

I plan to resume work at the zoo, connecting the remaining
floor drains and hanging the pipes for the water system.
If, however, work can be resumed at the hospital during the
week, I'll work toward completing that project.

**FIGURE 13-6**  An example of a progress report

the work accomplished in relation to each job is explained. If only one major job is
involved, various aspects of the job may be shown in the margin headings. See
Figure 13-7.

# RKG MANUFACTURING

Date:       September 4, 1987
  To:       Harold Sill, Plant Superintendent
From:       Wayne Jones, Safety Committee Chairman
Subject:    Safety Inspection for August 1987

The following violations were noted during the monthly safety inspection conducted on August 28, 1987:

I. Storage racks in the following areas were not bolted to the floor.
   A. In the warehouse
   B. In the northeast corner of the production area
   C. In the chemical-storage area
   D. In the paper-storage area

II. Oil spills were found
   A. On the entire floor in the storage room
   B. Around Press No. 2 in the pressroom

III. Increased ventilation is needed
   A. Above the engineering blueprint room
   B. Above the inert-gas and other arc welders

IV. Improved cleanup is needed
   A. On the entire plant floor, where scrap metal and paper were found
   B. In the storage area, where flammable materials were found

V. Hazardous conditions exist
   A. At the pit, where the gate was left open although the pit was not in use
   B. At the ovens, where some oven doors were left open
   C. In the warehouse, where cartons improperly stacked seemed ready to fall

VI. Conveyor-belt guards are missing
   A. On the cumulator conveyor
   B. On Press No. 6
   C. On both ends of the air-oven belts

FIGURE 13-7  An example of a periodic report

## Justification Reports

A justification report may also be called a recommendation report or a proposal, because it always implies or specifies some kind of change. More formal and explanatory than the suggestion-box method, the justification report is often merely another way of proposing the development of a new mechanism, the adoption of a new procedure, the purchase of additional equipment, or a change in company policy. The justification report is sometimes written as a memorandum or a form report but is more frequently written as a letter report. See Figure 13-8.

# LETTER REPORTS

A letter report, usually written by a person whose authority in a particular field is well established, is submitted as a formal business letter. As illustrated in Figure 13-8, the author's facts are rarely verified in a detailed explanation. The recipient accepts the facts because he or she believes the author is qualified to make evaluations and decisions. Some letter reports are summarizations abstracted from a detailed analytical or examination report that is filed for future reference. Figure 13-8 is a justification written in the form of a letter by an engineer to a plant superintendent. The report has a tone of authority suggesting that the information should be accepted without question.

Technicians probably use the formal letter report only if they are submitting information to an executive with whom they have no personal contact. Occasionally, however, technicians on an extended field trip write letter reports to their immediate supervisors. A letter report differs very little from a memorandum report except in format, as illustrated in Figure 13-8 and explained in the following comments.

1. The complete heading, inside address, and salutation used in business letters are also used for letter reports.
2. The subject of the report is usually stated. It may be placed on the same line as the salutation or two spaces below.
3. In the letter report, an introductory statement replaces the introduction used in a formal report.
4. The information is logically developed under appropriate margin headings.
5. The introductory statement in letter reports often has an informal tone characteristic of many business letters. The report information, however, is formal and objective.
6. A simple graphic aid, especially one presented as a table, may be used where it is appropriate.
7. A summary of information is not given.
8. If a recommendation is required, it is usually made in the introductory statement. However, if the author believes that the recommendation is more

**COMMERCIAL CONCRETE COMPANY**
**195 Collins Street, City, State**

January 4, 1986

Ms. Elizabeth Fife, Superintendent
Commercial Concrete Company
195 Collins Street
City, State Zip code

Dear Ms. Fife:

Subject:   Recommendation for Purchasing a Forklift

After we discussed the possibility of purchasing a forklift
to supplement the work performed by the front-end loader,
Plant B rented a forklift and used it for one year.  The
analysis given in this letter shows that purchasing a forklift will result in greater efficiency and economy at
Plant B.

Uses for a Forklift

A forklift can perform four types of jobs efficiently and
quickly:

1.  The forklift can be easily elevated and lowered to move
    merchandise from trucks and railroad cars and deliver
    it to the work area.
2.  The forklift can maneuver merchandise into hard-to-reach
    places.
3.  The forklift can quickly and easily move pelletized ma-
    terial from one location to another.  A year ago, a crew
    of six workers spent two days unloading a carload of
    this material each time a shipment arrived.  This year,
    a forklift and one worker accomplished the job in one
    day.  Ten carloads of pelletized material are received
    annually.
4.  The forklift can quickly handle and stack stoves.  Man-
    ual handling and stacking of stoves requires four work-
    ers six days a year.  With a forklift, three workers can
    complete this job in four days.

Financial Savings in Hours of Labor

Last year, the forklift saved 88 hours of labor at a cost
of $7 for each hour of labor to move pelletized material.
It saved 76 hours of labor at a cost of $7 for each hour of
labor to handle and stack stoves.

**FIGURE 13-8**  An example of a letter report

The total purchase price of the forklift is $10,000.  The
life expectancy is ten years.  At 6 1/2 percent interest,
the ownership cost of the forklift over five years is $2406
a year.  The net financial gain to the company during the
first five years is $4288.  After five years, all savings
in hours of labor, $6694, are profit.

Additional Advantages of the Forklift

A forklift on the premises, in addition to being used for
special work, can be used to rearrange merchandise or to
unload some deliveries.  It also reduces the risk of injury to workers attempting to move items manually that can
be more easily and safely moved with a forklift.

                                        Sincerely,

                                        James Rohn

                                        James Rohn, Engineer

                                        Plant B

**FIGURE 13-8** *(Concluded)*

effective after the argument has been stated, the recommendation may appear at the conclusion of the report.

9. A complimentary close and signature complete the letter report.

10. The report is single-spaced. Paragraphs are seldom indented, but a double space is used between paragraphs.

11. If a letter report extends beyond one page, each following page begins, 1 inch from the top, with the name of the recipient, the date, and the page number listed and single-spaced on the left-hand margin of the page. See the following example.

EXAMPLE

—Ms. Elizabeth Fife

—January 4, 1987

—page 2

# SUMMARY

1. Most organizations rely upon routine memorandums and form reports to communicate and coordinate ideas, developments, or useful information within a department or among various departments.

2. Memorandums have various uses and formats, as shown in Figures 13-1, 13-2, and 13-3.

3. Memorandum reports are the longest and most detailed memorandums. Nearly all stationery used for these reports has some preprinted text such as the words *Memorandum, Date, Subject, To,* and *From* printed near the top. Memorandum reports are usually explanations of processes, analyses, examinations, or mechanical defects. These reports substitute for formal reports only when the required text is brief and is written exclusively for personnel within the business or industry.

4. Form reports briefly explain ongoing projects such as routine laboratory procedures, routine progress on a project, or routine work-related activities. They are usually submitted on preprinted forms similar to those shown in Figure 13-4 (open text) or Figure 13-5 (short-answer text).

5. Like formal reports (Units 8 to 12), informal messages — whether written as instructions, comments, policies, or reports — are communications. Therefore, they convey to a reader the attitudes of the writer. For this reason, technicians who write them use standard English; consider the reader; use attractive format; and present all necessary text clearly, concisely, and logically. Writing style, however, is usually informal.

6. Letter reports, usually written by a person whose authority in a particular discipline is well established, can have content similar to that in memorandum reports. However, the page has no preprinted text and a more formal business-letter format is used. Letter reports, unlike memorandums, are frequently sent from one company to another. Guidelines in this unit suggest techniques that make letter reports effective.

---

UNIT **13**

## SUGGESTED ACTIVITIES

A. Discuss the following questions.
   1. How many pages are preferably used for a memorandum report?
   2. What parts of a formal report are omitted from a memorandum report?
   3. What is the primary difference between a memorandum and a letter report?
   4. Why is a memorandum or form report likely to be carelessly written?
   5. Distinguish between a progress report and a periodic report.
   6. What kind of spacing is used for memorandums, for form reports, for letter reports, and for formal reports?

7. Define objectivity in reports.

8. Define standard English.

9. Under what circumstances would a handwritten memorandum or form report be justified?

10. What kinds of graphic aids are used in memorandums and in letter reports? Where are graphic aids placed?

B. Using information from the formal reports written previously for this course or from another source, prepare a memorandum report. Use the background information, the purpose, and the major topics as guides for preparing the subject line and introductory statement in the memorandum. Use the rules for developing the body of a report as guides for preparing the text that follows the introductory statement.

C. Assume that you are on a two-month field trip to an affiliate overseas. Write a letter report explaining the progress being made on the project you were sent to initiate. Or write a letter report explaining the examination of a piece of equipment or the working conditions.

D. Try to obtain a memorandum or letter report from a library or a local industry. Using the guidelines given in this unit, write or present orally a constructive criticism of the report.

E. Try to obtain both open-text and short-answer form reports from an organization similar to one you plan to work for. The instructor can discuss any reports that differ from those explained in this unit.

# 14

# Proposals

## OBJECTIVES

After reading this unit, you should be able to do the following tasks:

- Describe the purpose and content of a proposal.
- Describe and list the types of proposal formats.
- Prepare and write a proposal.

## OVERVIEW

The subject matter of proposals can vary from an idea placed in a suggestion box to long, formal communications. Many proposals, especially those between large corporations and government or between one large corporation and another, are prepared by teams of experts. Therefore, a discussion of complete proposals does not belong in a basic study of technical writing. This unit is thus designed only to introduce technicians to some simpler kinds of proposals and some basic attitudes a writer may develop to help make all proposals effective.

# PROPOSALS DEFINED

In industry and public service organizations, a proposal is any communication designed to persuade one person or organization to accept some kind of service from another person or organization. Typical services are making an investigation, conducting research, solving some kind of problem, and constructing something. Some proposals are initiated by the person who wants to perform the work; others, by the person who wants the work done. They may be made from one person or department to another within an organization or from one organization to another.

In its simplest form, a proposal can be an oral request in which an employee asks a supervisor's permission to test a new product, to attend a seminar during company hours or at company expense, or to try a new procedure. A supervisor grants permission, or gives authorization, orally. Usually, however, a proposal and letter of request and authorization are written, if only to guarantee that these various transactions have taken place.

The proposal may be presented as a memorandum, as a letter, or as a formal paper. (Authorities argue whether a proposal should be classed as a report. By definition, a report tells about past performances; a proposal tells about something to be done in the future. Resolving the argument is not important. As explained later in "Suggestions for Preparing Effective Proposals," most of the attitudes and techniques used to prepare effective reports are also essential for effective proposals.)

# TYPES OF PROPOSAL FORMATS

As stated previously, a proposal may be presented as a memorandum, a letter, or a formal paper. The choice depends upon company policy, the reason for making the proposal, the wishes of the organization requesting the proposal, and the amount of data needed.

## Memorandum Proposals

Some companies, especially those who routinely make relatively short proposals for various kinds of construction or installation, adopt a standard memorandum similar to the one shown in Figure 14-1. This proposal is equivalent to placing a bid for a particular job. As shown in the preprinted text, the proposal includes specifications for the work and a complete cost for labor and materials. Also, most memorandum proposals sent outside the originating company are made in triplicate. Additional pages, also in triplicate, are attached when needed, and each page is numbered.

**Proposal** Page No. of Pages

**YOUR FIRM NAME HERE**
123 Main Street
YOUR TOWN, STATE AND ZIP
**Phone 123-4567**

| | | | |
|---|---|---|---|
| PROPOSAL SUBMITTED TO | | PHONE | DATE |
| STREET | | JOB NAME | |
| CITY, STATE AND ZIP CODE | | JOB LOCATION | |
| ARCHITECT | DATE OF PLANS | | JOB PHONE |

We hereby submit specifications and estimates for:

**We Propose** hereby to furnish material and labor — complete in accordance with above specifications, for the sum of:

__________ dollars ($ __________ ).

Payment to be made as follows:

All material is guaranteed to be as specified. All work to be completed in a workmanlike manner according to standard practices. Any alteration or deviation from above specifications involving extra costs will be executed only upon written orders, and will become an extra charge over and above the estimate. All agreements contingent upon strikes, accidents or delays beyond our control. Owner to carry fire, tornado and other necessary insurance. Our workers are fully covered by Workmen's Compensation Insurance.

Authorized Signature __________

Note: This proposal may be withdrawn by us if not accepted within __________ days.

**Acceptance of Proposal** — The above prices, specifications and conditions are satisfactory and are hereby accepted. You are authorized to do the work as specified. Payment will be made as outlined above.

Signature __________

Date of Acceptance: __________

Signature __________



**FIGURE 14-1** An example of a memorandum proposal (Source: New England Business Service (NEBS), Groton, Massachusetts)

A proposal prepared for someone within an organization, such as one seeking permission to make changes of various kinds within a department, is usually presented as a memorandum similar to the one in Figure 13-3. As shown, headings identify the various topics detailed in the memorandum.

## Letter Proposals

Letter proposals, like letter reports (Figure 13-8), are written when a communication more formal than a memorandum seems appropriate. Unlike letter reports, letter proposals may be sent not only to personnel within an organization but also to another organization. For example, a foreman may use a letter to send a proposal to a chief executive. A letter proposal to another organization is used when necessary data can be presented in no more than three pages but a memorandum seems too impersonal or informal.

The tone of the letter proposal is conversational; occasionally, the introduction even refers to a mutual interest or acquaintance. Headings introduce the major topics and possibly the subtopics.

In a proposal requesting work, the proposer tells why he or she is interested in the project, in what specific ways the proposer is qualified to perform the work, and what procedures will be used to successfully complete the work. Total cost is given in the letter if stating only the total cost is adequate. Otherwise, an itemized cost sheet is attached to or enclosed with the letter. See Figure 14-2.

In a proposal requesting financial aid or other kinds of support, the proposer tells what service the organization making the request performs; identifies the plans, goals, or immediate needs that require support; specifically requests support; and briefly explains how the support will benefit society and, if possible, the supporter.

One or two closing paragraphs are included in all letter proposals. The text and tone of the letter and the relationship existing between the two parties involved in the correspondence help determine closing remarks. Often, the proposer requests an early response and specifies a deadline. He or she may also express appreciation or, as in Figure 14-2, ensure that work can begin now or on a specified date.

## Formal Proposals

As mentioned earlier, formal proposals, especially those requesting work, can be long — one hundred or more pages. They can also contain numerous graphic aids showing available equipment, statistical data, and charts of various kinds. However, many formal proposals are much shorter — perhaps four to seven pages. Formal proposals, regardless of length or kind, are arranged in a format similar to that of a formal report.

COMPANY NAME
Street Address
City, State   Zip code
May 15, 1986


Mr. Harold Wheeler
Atlas Coal Processing Company
Street address
City, State   Zip code

Dear Mr. Wheeler:

Subject:      Analyzing the Cause of Excessive Moisture in
              the Product Dryer Feed

Yes, as outlined below, Consultant Engineers, Inc., does
have all the personnel and facilities needed to analyze the
cause of excessive moisture in your product dryer feed.   In
fact, we have done similar analyses for several companies
in the Intermountain West.   Their names are available upon
request.

Personnel and Facilities

Our staff includes all disciplines required for plant engineering, particularly in materials handling, pressure
piping, and instrumentation systems.  In addition, we have
experienced project management personnel skilled in all
techniques required to start projects, maintain schedules,
monitor and control project costs, and put completed facilities into operation.

Members of our purchasing staff have immediate access to
suppliers, vendors, and fabricators in the Intermountain
area.   They also have access to suppliers and manufacturers
throughout the United States.

Our corporate staff consists of 20 inspectors located geo-
graphically over the United States.   The one located as
close to you as Utah can maintain quality control at minimum cost to you.

Suggested Procedures

Several methods for reduction of excessive moisture should
be tested to determine the most effective one.   These methods are listed below.

**FIGURE 14-2**  An example of a letter proposal

Mr. Harold Wheeler
May 15, 1986
page 2

1. Introduce additional heat into the dryer.

2. Recycle a portion of the hot, dry product from the dryer discharge screw conveyor to the dryer feed screw conveyor through one of the following:

   a. A tubular chain conveyor

   b. A "Redler" type conveyor

   c. Other en masse type conveyors

   d. A pneumatic conveyor

3. Recycle some of the dry dust product from the bag-house discharge chute to the dryer feed screw.

4. Change the centrifuge or slurry system if either is found to be defective.

<u>Cost</u>

The cost for these test procedures, to be determined when the tests have been completed, is based upon the hourly rate shown in Schedule A, attached.

As soon as we receive your authorization, we can begin the tests. Once they are completed, we are also interested in submitting a proposal for effecting the removal of excessive moisture, using the most efficient method indicated by the tests.

Sincerely,

Dennis Blake

Dennis Blake
Chief Mechanical Engineer
Consultant Engineers, Inc.

**FIGURE 14-2** *(Concluded)*

The title of the proposal and other necessary information similar to that in Figure 7–3 are preferably arranged on a title page. The rules for creating the title duplicate those used to prepare report titles, as shown in the following example.

**EXAMPLE**

**A PROPOSAL FROM NATIONAL CONSTRUCTION, INC.,**

**FOR THE CONSTRUCTION OF CONDOMINIUMS**

**AT 1970 CHERRY LANE, CITY, STATE**

The proposal introduction consists of only two parts from the basic outline used for reports: the background and the list of major topics. The background is determined primarily from previous correspondence but should also arouse the reader's interest in the proposal and establish rapport between the reader and the proposer. The major topics may be listed or briefly explained. No headings are used within the introduction.

The following guidelines help proposal authors organize the remaining text for a work request.

- The work the company intends to accomplish is explained in detail. In this discussion, enough information is presented to convince the recipient that the proposer fully understands the project.

- Accurate descriptions are given of the facilities and capabilities within or available to the proposing company that qualify the company to perform the work.

- Clear, specific, and complete technical information is given to show how the work will be done.

- Details of the work schedule are presented, including dates for beginning and completing the project.

- The cost of the project is given, usually in an itemized list preceded by a lead-in sentence.

- Appropriate graphic aids are incorporated any time their use clarifies or emphasizes important text.

- Each major topic and possibly subtopic in the proposal is preceded by the appropriate headings, as explained in Unit 7 and shown in Figure 7-4. A summary is not likely to be used. The companies involved decide whether the proposal is single- or double-spaced.

- A short letter of transmittal (Unit 16) is usually sent with or attached to the proposal. A table of contents (Unit 16) is also recommended for proposals longer than five pages.

Organizations requesting funds or services that may or may not be available usually write proposal letters. If funds such as tax dollars or grants are available but one organization is competing with others for them, a formal proposal is often written.

Good judgment is essential in organizing effective proposals that request funds. For example, in developing the subjects to be discussed in the proposal, an author needs to evaluate how much information to give. A first request needs more than a second one does. Also, the text must continually build the reader's empathy, understanding, and desire to contribute. The following guidelines help a proposal writer organize the text of a request for funds.

- The introduction, which differs from that for formal reports but is similar to that for an oral report (Unit 15), is written to arouse reader interest and to gain approval of the request.

- The services the organization performs and the social benefits of these services are explained. If the organization has previously received funds from the proposal recipient, such as tax dollars for medical research, the author explains all developments resulting from the use of those funds.

- The situations or conditions that create the request, such as additional research, new facilities, replacement of worn-out or outdated equipment, or loss of previous financial support, are explained. Any comments that justify these needs are incorporated.

- Benefits that can or will result from new funding are explained. For example, additional funds may permit studies leading to more durable road bases, more effective environmental control, or new disease-prevention procedures. Specific data showing the feasibility of these studies is essential.

- The request for funds is made specifically and directly. The request follows one or more statements that relate the services and needs previously discussed to the request. In addition, if the request is for specific purchases or new employees, an itemized list reflecting the fairness of the request is prepared.

- A brief summary recapitulating the report text is recommended.

- The organization submitting the request decides whether graphic aids contribute to or detract from the text.

- A short letter of transmittal (Unit 16) is attached to or submitted with the report.

- A table of contents (Unit 16) is recommended for proposals longer than five pages.

- The organization submitting the proposal decides whether single or double spacing is used.

# SUGGESTIONS FOR WRITING EFFECTIVE PROPOSALS

As implied earlier, a proposal is a sales communication. It is written to convince someone that a particular company has the best service available for the work to be done or that a particular service organization is the one that should receive available funds.

Nearly all organizations requesting funds rely upon compliance with the request to survive. Similarly, many industries such as construction, defense, or consulting firms depend almost exclusively upon contracts to stay in business. These contracts are usually issued after proposals have been presented. Thus, when a proposal is rejected, costs for time and energy used in the proposal's preparation are never recovered. More importantly, however, the rejection can hinder the flow of new work that may be essential for the company's continuous productivity. For these reasons, a person writing or helping to write a proposal should recognize the value of an effective proposal and also become familiar with the following suggestions.

1. The proposal writer should understand the reader and anticipate the reader's needs; the writer wants to neither bore the reader with excess data nor leave the reader with unanswered questions. The writer also organizes the data in a way that has the greatest sales appeal. (Sales appeal in proposals is accomplished through an honest, confident presentation of verifiable information, not through exaggeration or promises that cannot be fulfilled.)
2. A proposal writer should possess confidence and the ability to use good judgment in making decisions. Confidence permits the author to write authoritatively and thus convincingly. Judgment helps him or her determine how much data is needed and how it should be stated. For example, one executive who works with a team of experts submitting proposals to various departments of the federal government learned from experience that a conversational approach is sometimes more effective than an impersonal one. The executive also learned that different recipients respond differently to identical organization of information, writing style, and format. Therefore, this person is convinced that proposals are more successful when judgment is used to evaluate all circumstances before each proposal is prepared.
3. Two important questions to be considered by every proposal writer are, "What is our company's strongest selling point?" and "In what important ways do our products, facilities, and personnel excel?" These questions are appropriate for an introduction. Answers to them can be incorporated into the general text. However, the questions and answers should be implied, not specifically stated. Preferably, they become evident through positive, confident statements related to a company's reputation, standards, and capabilities.

4. Substandard English, incorrect spelling, and illegible handwriting, even in informal memorandum proposals, are never justified. The slang terms and jargon often acceptable to fellow employees when spoken never have the same degree of acceptance when written.

5. All proposal writers should apply the various guidelines that result in accuracy, completeness, clarity, and conciseness. Clarity and conciseness are particularly important; a bored or confused reader is not likely to be a willing buyer.

6. Graphic aids can be important in proposals for the same reasons they are important in reports. They can be used in memorandum and letter proposals as well as in formal proposals. The guidelines discussed in Unit 5 apply.

# SUMMARY

1. A proposal is a sales communication submitted to convince someone that an organization, a department, or a person can provide a particular service. The possibilities and needs for proposals are extensive. In fact, the acceptance or rejection of proposals often determines whether an organization will survive.

2. Proposals may be short, simple suggestions placed in a suggestion box or expressed orally. Other proposals are frequently so long and complex that a team of experts prepares them.

3. This unit is not designed to explore all kinds of proposals. It is written primarily to alert technicians to the prevalence of proposals, to their importance, and to the likelihood that a high percentage of employees can expect to write or help write proposals.

4. This unit divides proposals into three general types: the memorandum proposal, the letter proposal, and the formal proposal.

5. Memorandum proposals are usually submitted on a standard, preprinted form (Figure 14–1).

6. Letter proposals are written in letter-report format, as in Figure 13–8. The tone of letter-proposal introductions is conversational (see Unit 18). However, the specific proposal within the letter is presented in the objective style used for formal reports. Headings and graphic aids are used when needed.

7. Formal proposals, as explained on pages 162 to 166, follow the outline, the format, and the objective style used for formal reports.

# SUGGESTED ACTIVITIES

A. By asking a representative organization or by reviewing this unit, decide what kind of proposal an organization you plan to work for is likely to use. For example, construction firms may use only memorandums; consulting firms probably use formal proposals; manufacturing firms are likely to use either letter or formal proposals. These firms are all seeking some kind of work. Public service proposals are usually written for a different reason; they most frequently request additional funds, facilities, or equipment. Letter or formal proposals are used. After selecting an appropriate format, write a proposal.

   Note: If a formal proposal including a letter of transmittal and table of contents seems to be the best choice, two or more students may want to work together on a single proposal.

B. If possible, visit an organization that interests you, and interview a supervisor about the kind of proposal writing used there. Before going, use the information in this unit to prepare a set of questions to be asked. Also, be willing to listen to any suggestions the interviewee may offer concerning techniques for making proposals effective. Keep accurate notes; then, in an oral report, discuss the interview in the classroom. (Guidelines for presenting an oral report are given in the next unit.)

# 15

# Oral Reports

## OBJECTIVES

After reading this unit, you should be able to do the following tasks:

- Plan an oral report.
- Describe the techniques for building confidence as a public speaker.
- Discuss the guidelines for relating to an audience.
- Integrate graphic aids into an oral report.
- Prepare for answering questions after the presentation of an oral report.

## OVERVIEW

The objective for giving an oral report is identical to that for preparing a written report — to transmit information. In conferences, business meetings, seminars, and professional groups, technicians may be principal speakers or contribute to discussions. For example, a designer of a new mechanism explains its operation to other employees or to potential buyers. A machinist describes the procedures or problems involved in a specialized machining procedure. A technician back from a field trip reports his or her work-related experiences to department members or company executives. Developing the art of speaking effectively benefits every technician because speaking, like writing, is necessary in all business and industrial communication.

Many excellent courses and textbooks teach principles of public speaking. This unit is limited to explaining a few techniques for presenting an oral report effectively through planning the report, building confidence, relating to the audience, using graphic aids, and answering questions.

# PLANNING THE ORAL REPORT

Speakers use one of four methods to present an oral report:

1. They read it.
2. They memorize it.
3. They talk from notes.
4. They talk from an outline.

A person who reads or memorizes a report does not consider the audience as part of the performance. The speaker's objective can only be to talk about the subject. A technical speech that is read or memorized is insulting to a group of people interested enough to attend the meeting. Far preferable to a memorized speech or one read from a written report is the one prepared carefully enough that it can be discussed from written notes or an outline. Only by one of these latter two methods is a speaker truly able to relate to the audience.

The advantages for using an outline for an oral report are identical to the advantages for using an outline for a written report. The outline organizes the information; organizes the speaker's thoughts; establishes a beginning and an end for the speech; and guarantees that the information used is pertinent, specific, and complete. These advantages suggest that every oral report should be outlined. Once the outline is prepared and rehearsed, the speaker usually prefers the outline to notes as a guide while speaking. In fact, a speaker who, while speaking, displays an easy-to-read, attractive outline on a poster or video screen encourages an audience to move along with him or her.

The outline for an oral report is identical to the one prepared for a similar written report. However, a speech differs from written text in several ways, as explained in the following guidelines.

- The language in a speech is informal. Frequent references to *I, we,* and *you* help establish communication between the speaker and the audience. Although correct grammar and sentence structure are still necessary, the choice and arrangement of words are more conversational.

- The background information in a speech is planned to arouse the interest of the audience and to establish rapport between the audience and the speaker. Jokes and other humorous comments unrelated to the report should be avoided. However, a speech can be introduced with anecdotes, personal experiences related to the subject, a challenging question, or an interesting graphic aid. Technicians who have studied speech may be aware of other interesting techniques for beginning a technical speech.

- The statement of purpose should not be omitted in an oral report. The listener, like the reader, wants to know what will be accomplished. The purpose is usually presented less briefly and more informally in a speech than in a written report but is always clear and specific.

- Necessary definitions are made in the introduction. A reference to the original definition may be required when the defined word is used in the discussion. Difficult technical terms may need to be written on a blackboard or large piece of paper where the audience can see as well as hear them.

- Classes in dramatic expression emphasize that important ideas need to be presented to an audience three times, each time in a different but meaningful way. A speaker who realizes that repetition is an effective method of emphasis will not hesitate to state all major topics and clarifying statements about them in the introduction, in the body of the speech, and in the summary.

- Novice speakers sometimes hesitate to summarize an oral report because they fear repetition. Skillful repetition, however, does not sound repetitious. It emphasizes important concepts. The summary of an oral report should be planned to give the audience a condensed but unified mental picture of the process, the problem, or the analysis discussed in the report. It also ensures that the purpose of the report has been accomplished.

# BUILDING CONFIDENCE

Confidence is the faith people have in themselves as a result of an honest evaluation of their abilities. Public speakers develop self-confidence through the study of and practice in public speaking. Technicians develop confidence in their ability to interest audiences in technical reports through the following techniques: knowledge of the subject, a positive attitude, preparation, and rehearsal.

## Knowledge of the Subject

Technicians are not usually asked to speak about technical subjects unless they are qualified. Nevertheless, technicians may understand a topic thoroughly and still overlook details needed to make the subjects meaningful to uninformed audiences. Confidence in their ability to transmit knowledge to others is increased when technicians relate their subjects to the needs of their audiences.

## A Positive Attitude

Effective speakers not only understand their subjects but also believe in them. They believe their audience wants to learn about the subjects. Effective speakers also have a positive attitude toward oral reports. A positive attitude toward the subject, the audience, and the speech contributes to an effective presentation, and the audience becomes more receptive because of it.

## Preparation

Some speaking groups and speech instructors advocate impromptu, or unprepared, speeches. Such speeches are designed to help a person overcome the fear of speaking. They are not designed for the communication of technical information. The importance of thorough preparation of an oral report cannot be overemphasized. Speakers who control their information by organizing their ideas and understanding their data can speak confidently and convincingly to inquisitive audiences.

## Rehearsal

Rehearsal of an oral report is essential for many reasons. Even skilled speakers need to review their topics until they are confident of the following points:

1. Their speeches are well organized.
2. The meaning will be clear to uninformed audiences.
3. They will not exceed the time limit.
4. Graphic aids are incorporated effectively.
5. The presentations will be interesting.

They also need to practice aloud to determine whether they speak too rapidly or too slowly, whether their important ideas are being stressed, and whether their tone is convincing but not demanding. A person who is satisfied with the performance during the practice session is confident when he or she faces an audience.

# RELATING TO THE AUDIENCE

An oral presentation of technical information is not merely an explanation of a mechanism, a process, an analysis, or problems relating to medicine, fire protection, or food preparation. It is communication with a group of people.

The listeners for whom an oral technical report is prepared are not expecting humor or superficial explanations. They are seeking information. The speaker, therefore, is expected to be serious but not boring, knowledgeable but not showy. Effective communication between a speaker and the audience usually develops if a speaker considers the following points.

1. Effective speakers develop an interest in their audience. If they genuinely want their listeners to understand the topic being discussed, they will win audience support.
2. Good speakers are confident about their subjects and their ability to explain the topics, but they are unassuming in their manner. The listeners want to feel that the speaker is talking with them, not at them.

3. Good speakers are human. They watch their audiences as they talk and are not afraid to display warmth and friendship in manner and voice.
4. Good speakers sense when their listeners are losing interest and attempt to restore it. They change their tone, ask a rhetorical question, cite an interesting anecdote related to their subjects, change their tempo, or bypass details that may be dull or confusing. In contrast, they simplify a discussion that seems too technical or add detail to one that seems too sketchy.
5. Good speakers cover their subjects and complete their speeches within the designated time.

Some mannerisms or speech habits that may distract, irritate, or bore an audience and should be avoided are listed here.

—Unusual clothing or jewelry that draws attention to itself.
—Rhythmic noise such as playing with coins or keys or tapping on a tabletop or speaker's stand.
—A monotonous tone in the speaker's voice. Although few untrained speakers have perfect tone control, an enthusiastic speaker is rarely monotonous.
—The use of *and-uhs* and similar meaningless words between sentences. This tendency may be curbed through preparation and rehearsals.
—Words that are not expressed fully and distinctly and are pronounced incorrectly. Practicing aloud may help overcome this habit.
—A speaker who paces nervously. However, one who moves purposely or gestures spontaneously because of enthusiasm keeps the audience interested. When a microphone is used, a speaker should remember to stay within its range.

# USING GRAPHIC AIDS

Speakers who appeal to the sense of sight as well as to the sense of hearing improve their chances of being understood. Carefully chosen graphic aids that are skillfully used create the visual support for the spoken text. The following suggestions offer techniques for preparing graphic aids and using them effectively.

1. The types of graphic aids that may be used are almost unlimited. Flow-charts, graphs, tables, maps, slides, pictures, and simple diagrams are a few examples. In addition, one of the best graphic aids is the object being discussed in the report. For example, one technician effectively demonstrated how electricians climb poles by showing, during the discussion, how pole-climbing equipment is worn and used.
2. The choice of graphic aids depends upon the purpose of the report, the size and kind of audience, and the equipment available in the lecture room. For example, if the room has no chalkboard, a speaker cannot use chalkboard

illustrations. If electric outlets have not been installed or do not function properly, electrically controlled graphic aids must be eliminated. The speaker is obligated, therefore, to determine the limitations of a lecture room before selecting graphic aids.

3. Chalkboard illustrations are used only when the speaker can write and illustrate neatly and clearly. The speaker, after considering the subject and the audience, should determine whether to prepare chalkboard illustrations before or during the speech. Electrical equipment, experiments, and other graphic aids that require time to prepare should be set up and tested before the speech begins.

4. Mechanical devices such as slide projectors can be overused. An effective speaker selects slides carefully so that pictures not directly related to the purpose of the report are omitted. The selected slides are arranged in the correct order for presentation and are not separated from one another by blank compartments. Saving a series of slides to be flashed before the audience at the end of a speech is rarely justified. This procedure is comparable to placing important graphic aids in an appendix.

5. All graphic aids must be clear, neat, interesting, and informative. Colored illustrations are usually more appealing than black-and-white ones. The details in a graphic aid must be visible to everyone in the room. As in written reports, graphic aids must be introduced and their relationship to the report explained.

6. Graphic aids are not substitutes for a well-planned speech. As aids, they are secondary to the discussion.

7. A graphic aid, if it creates interest and is appropriate, can be effectively used at the beginning of an oral report.

8. If the audience is small, some speakers permit samples, small pictures, and similar graphic aids to be passed from one person to another during the speech. The speaker must decide whether this procedure will aid communication or be distracting.

# ANSWERING QUESTIONS

An oral report is usually followed by questions from the audience. Experienced speakers are adept in conducting question-and-answer sessions. Inexperienced speakers may find the following comments useful.

1. Speakers should always learn as much as possible about their subjects before giving oral reports.

2. Speakers can anticipate some questions by imagining themselves as a member of the audience.

3. Speakers must be patient and understanding while people are asking questions. Sometimes, the people are nervous and, therefore, fail to make their

questions clear. Rather than attempt to answer a question the speaker does not understand, a speaker asks that the question be repeated or restates the question as he or she understands it and asks whether that interpretation is correct.

4. Speakers who admit they cannot answer a particular question receive more audience approval than ones who try to bluff.

5. All questions and answers should be audible to everyone in the room. If the question is indistinct to others, the speaker repeats the question before answering it. While answering, he or she addresses the entire audience, not just the interrogator.

6. When an audience becomes restless or the time limit is near, the speaker is responsible for ending the meeting, usually by asking for a final question.

# SUMMARY

1. Technicians are frequently expected to discuss or report work-related information orally to one person or to an audience. Therefore, oral communication is as important as written communication for job success.

2. This unit is not a substitute for public-speaking courses. The information is confined to planning the oral report, building confidence, relating to the audience, using graphic aids, and answering questions from the audience.

3. Maintaining audience interest cannot be overemphasized. Unless people listen, the speaker's efforts are wasted. Every aspect of a speaker's preparation should, therefore, be directed toward relating to the audience. The guidelines on pages 173 to 174 help build this relationship.

4. Several suggestions for coordinating appropriate graphic aids with oral presentations are offered in this unit. If carefully selected and presented, graphic aids hold audience interest and thus contribute to the effectiveness of oral reports.

5. Because the content of an oral report is usually unfamiliar to many in an audience, opportunities for them to ask questions and receive answers are appreciated.

UNIT **15**

# SUGGESTED ACTIVITIES

A. In the following list of words, letters frequently slurred or mispronounced are underlined. Check the dictionary for correct pronunciation if necessary; then, while concentrating on the underlined letters, say the words aloud. This practice helps eliminate slurred speech.

1.  thermometer
2.  temperature
3.  perpendicular
4.  accurate
5.  recognize
6.  quantitative
7.  intramolecular
8.  interject
9.  photometer
10. calculate

11. for instance
12. interpretation
13. environmental
14. infrared
15. coronary
16. something
17. climactic
18. satisfactory
19. introduce
20. hesitate

**B.** Assuming that you are speaking to an audience, read the following sentences aloud. Speak slowly enough that each word is distinct. Change your voice to place emphasis on important words. Pause briefly where punctuation marks indicate a transition in thought.

1. In quality control, two types of analysis are used: chemical analysis and physical analysis.

2. When an internal gear is meshed with a pinion, the gear ratios are similar to those for an external gear mesh.

3. Can a radio-frequency oscillator be used to determine high concentrations of alkali in aqueous solutions?

4. A change in temperature can seriously affect the results of the experiment.

5. James Watt, the famous Scottish inventor, used steam to heat his office in 1784.

6. To eliminate the bad odors and tastes in water, the chemist added activated charcoal before the water was passed through a bed of floc, sand, and gravel.

7. Since the molecules are extremely small, they can move through pores that are invisible even under a microscope.

8. Perfume evaporates slowly; the escaping gas announces the presence of the wearer.

9. If the tube is heated, the mercury in the tube moves upward; if the tube is cooled, the mercury moves downward.

10. A Wheatstone bridge circuit is composed of two delta networks having one common side.

**C.** Attend a meeting in which an oral report is being given or attend a lecture concerning a technical subject. Lectures of this kind are given for engineering or other technical groups on most college campuses. Make a critical analysis of the speech.

   1. Was the speech interesting? If so, what techniques did the speaker use to arouse interest? If not, could the speaker have done something to increase interest?
   2. Was the speech informative? Justify your answer.
   3. Name some specific ways in which you think the speech could have been improved.

D. Organize a technical report to be given in a classroom. Make your introduction interesting, but include all information needed to clarify the purpose of the report and the major topics. Plan to use some graphic aids. During rehearsals, arrange the graphic aids in an order that will permit them to be presented without confusion or loss of time. Prepare a summary that reemphasizes the important ideas presented in the report.

E. Give the oral report in the classroom. After the talk is concluded, accept constructive criticism from the audience and use it to improve your next oral report.

F. Create opportunities to speak frequently before a group or as a member of a panel or seminar. Recording a minute or two of the speech will aid in self-analysis.

# SUGGESTED SUPPLEMENTARY READING

Damerst, William A. *Clear Technical Reports.* New York: Harcourt Brace Jovanovich, Inc., 1972. Pages 220 to 240 show an example of a single analysis and explain how the analysis is converted to a proposal.

Houp, Kenneth W., and Thomas E. Pearsall. *Reporting Technical Information.* Proposals, pages 350 to 367, and oral reports, pages 416 to 441. New York: Macmillan Publishing Company, Inc., 1984.

Lannon, John M. *Technical Writing.* Boston: Little, Brown and Company, 1985. The following kinds of reports are discussed and illustrated: physical description, pages 181 to 189; analysis, pages 483 to 505; and proposals, pages 420 to 458. Memorandums are discussed and illustrated on pages 399 to 415.

Markel, Michael H. *Technical Writing: Situations and Strategies.* New York: St. Martin's Press, Inc., 1984. The author explains and illustrates two physical-description reports, pages 168 to 177; a process report, pages 186 to 189; an instruction guide in the imperative mood, page 195; and memorandums, pages 263 to 285.

Mills, Gordon H., and John A. Walter. *Technical Writing.* 4th ed. New York: Holt, Rinehart and Winston, Inc., 1978. In a one-page chapter (2), the authors tell what a report should do.

Rathbone, Robert R. *Communicating Technical Information.* Reading, Mass.: Addison-Wesley Publishing Company, Inc., 1966. In Chapter 5, "The Inadequate Introduction," pages 33 to 40, the author discusses and illustrates report introductions. On pages 78 and 79, he suggests procedures for organizing technical description.

Roundy, Nancy, with David Mair. *Strategies for Technical Communication.* Boston: Little, Brown and Company, 1985. On pages 137 to 149, an outline and physical description of a hand auger are given. Short proposals are discussed on pages 257 to 260; progress reports, pages 271 to 283; memorandums, pages 363 to 364.

Sternberg, Patricia. *Speak Up! A Guide to Public Speaking.* New York: Julian Messner, 1984. This is a short, easy-to-read book that offers practical guides for developing an effective speaking voice, overcoming stage fright, organizing the speech, and giving the speech.

## Proposal References

The following two books are devoted entirely to proposal writing. The first concerns company proposals for government contracts. The second concerns social service organizations seeking donations or other funds.

Holtz, Herman, and Terry Schmidt. *The Winning Proposal . . . How to Write It.* New York, St. Louis, San Francisco: McGraw-Hill Book Company, 1981.

Kalish, Susan Ezell, and others, eds. *The Proposal Writer's Swipe File.* Washington, D.C.: Taft Corporation, 1983.

# Report
# Supplements

# 16

# Attachments to Formal Reports

## OBJECTIVES

After reading this unit, you should be able to do the following tasks:

- Prepare and write letters of transmittal.
- Prepare a table of contents.
- Prepare tables of tables and figures.
- Prepare and write abstracts.
- Prepare and write glossaries.
- Prepare and write appendixes.
- Organize report supplements.

## OVERVIEW

Special information is often attached to formal reports but not to memorandums or letter reports. The attachments discussed in this unit include letters of transmittal, tables of contents, tables of tables and figures, abstracts, glossaries, and appendixes. The placement of these attachments in relation to the completed report is discussed at the end of the unit.

# LETTERS OF TRANSMITTAL

A letter of transmittal, sometimes called a cover letter, is a communication used to present a report to a reader. A letter of transmittal may be attached to the report or forwarded in a separate envelope. It is always written when a report requested by one company is prepared by another company. For example, accounting firms often receive requests to prepare analytical reports for service organizations or industrial firms. Land-development companies investigate property for potential investors and submit investigation reports. Such requests are usually acknowledged immediately. The letter of transmittal is sent later with the completed report.

The need for letters of transmittal within a company is determined more by the size of the company, formality among the employees, and company policy than by standardized rules. However, any requested report not delivered personally should be accompanied by an explanatory letter or memorandum.

Letters of transmittal may vary in content and length. Often, a short paragraph referring to the date on which the request was made and specifying the title or purpose of the report is all the information needed, as indicated in the following example.

**EXAMPLE**     **Attached is the report you requested May 5, 1980, concerning the examination of the two drill bits operating at Monarch Mines.**

A statement similar to the one in the example, frequently written as a memorandum, is attached to a report that is handed to a supervisor or placed on the supervisor's desk. When written as a formal letter, it may also be used for a report to which an abstract, a condensation of a report, is attached.

A longer, more informative letter of transmittal, usually written as a business letter, is illustrated in Figure 16-1. The following guidelines related to Figure 16-1 explain the kind of subject matter used in most informative letters of transmittal.

- If the letter is replying to a request, the letter begins with a reference to the request and the date on which the request was made. The opening paragraph also tells whether the report is attached, enclosed, or being mailed in a separate envelope. If a request was not made, the opening paragraph may state the purpose of the report.

- If a specific report title is part of a request, that title is stated in the letter of transmittal and is used for the report. If only the purpose of the report is specified in the request, the letter need only restate the purpose, which usually is similar to the title of the report. An exact title is written entirely in capital

January 11, 1987

Mr. J.C. Wiles, Manager
Purchasing Department
Equipment Manufacturing Company
185 Augusta Road
City, State Zip code

Dear Mr. Wiles:

The report you requested January 5, 1987, concerning an
analysis of Microscope A and Microscope B for use in stereophotography of metal surfaces is attached to this letter.
The analysis of each microscope, based upon operation techniques, picture quality, and costs, shows that Microscope B
should be purchased.

As the photographs in the report illustrate, the rigid
attachment of a 35-millimeter camera to Microscope B produces consistently superior pictures to those taken through
Microscope A.  The swing-arm attachment on Microscope A
makes critical focusing difficult and uncertain.  A one-
step enlargement process achieves the desired magnification
ratio of 30 to 1 when Microscope B is used.  Two steps are
needed when Microscope A is used.

Although Microscope B costs $1200 more than Microscope A,
the higher-quality pictures and timesaving procedures resulting from the use of Microscope B justify the increased
cost.

The cooperation given by Mr. A.C. Bolin at Commercial Photographers, Incorporated, made an objective analysis of the
two microscopes possible.

If, after reviewing the report, you have questions concerning my recommendation, may I have an opportunity to discuss
it further with you?

Sincerely,

Ross Haile

Ross Haile, Illustrator
Editing Department

FIGURE 16-1  An example of a letter of transmittal

letters. The reference to the request, the date on which the request was made, and the purpose are frequently written in one opening paragraph.

- Subsequent paragraphs discuss important concepts, conclusions that result from the concepts, information that the writer considers important or interesting to the reader, or ideas the writer wants to emphasize. The information given in Figure 16-1 was presented to justify a recommendation.

- A recommendation given in the report is usually included in the letter of transmittal, especially if a request calls for a recommendation. The recommendation may follow the statement of purpose, as shown in Figure 16-1, or may follow the discussion of the report text. The position is dependent upon whether the reader will readily accept the recommendation.

- People who have assisted in preparing the report are given credit for their help in the letter of transmittal.

- The closing paragraph of the letter often specifies that the report writer is willing, if necessary, to discuss the report personally or to make further studies. However, the complimentary close (the words that come before the signature) may immediately follow the discussion part of the report. A closing statement similar to "I hope this report is satisfactory" should be omitted. The statement suggests doubt or apology. It is also becoming a cliché.

## TABLES OF CONTENTS

A table of contents has little value when a report is less than six pages long. For a six- to ten-page report, a table of contents is optional. For a report longer than ten pages, a table of contents is important. A table of contents, which is a complete, outlined guide to the report text, lists and gives page numbers for the headings used in the report and for the attached supplements. Figure 16-2 shows how the supplements discussed in this unit and the major and first-level headings for a report similar to the one in Figure 10-2 are arranged in a table of contents.

The following guidelines clarify the information included in the example and explain variations that may be made in the format.

- *Table of contents*, written entirely in capital letters and centered horizontally, is placed 2 inches from the top of the page. Four spaces separate the title from the first entry. *Page* is written two spaces below the title inside the right-hand margin.

TABLE OF CONTENTS



iii

**FIGURE 16-2** An example of a table of contents

- All letters in major headings and names of supplements may be capitalized. If a writer prefers, only the first letter of all important words is capitalized. Capitalizing all letters helps distinguish major headings from subheadings.

- The supplement pages that precede the report — the letter of transmittal, the table of tables, the table of figures, and the abstract — are given small

Roman numerals beginning with *i*. The title page is i, but the number is not placed on the title page and the title page is not included in the table of contents. Therefore, the table of contents begins with the letter of transmittal, page ii.

- The major headings and the subheadings used in the report are the major headings and subheadings used in the table of contents. All headings given in the table of contents must be used in the report, but all headings in the report do not necessarily appear in the table of contents. For example, paragraph headings are usually omitted.

- Roman numerals often precede the major report headings, as shown in Figure 16–2. This method of listing is optional. If the Roman numerals are omitted, all major headings and supplements have identical left-hand margins. The subheadings are indented five spaces, as shown in the following example.

**EXAMPLE**

**ABSTRACT**
**INTRODUCTION**
    **Purpose of the Report**
    **Definitions**
    **Sources of Data**
    **Topics of Discussion**

- Double spacing is used above and below major headings and supplement titles. Single spacing is used between all subheadings. If Roman numerals are used for major headings, as shown in Figure 16–2, the first-level subheadings begin under the first letter of the major heading; second-level subheadings, if used, are indented five spaces. See the following example.

**EXAMPLE**

**IV. ELEMENTS OF SUCCESSFUL PRODUCT DEVELOPMENT**
    **Systematic Product Development Program**
        **Initial Screening**
        **Business Evaluation**
    **Systematic Marketing Program**
        **Market Needs**
        **Market Potential**

- The right-hand margin is even. If Roman numerals are used, the left-hand margin varies to accommodate two- and three-letter Roman numerals.

# TABLES OF TABLES AND FIGURES

A report long enough to justify a table of contents usually contains enough graphic aids to justify a table of tables, a table of figures, or both. A short list of tables and figures may be placed below a short table of contents if all information can be placed on one page. Usually, however, separate tables are prepared. Occasionally, a single table includes both tables and figures when only two or three of each are used in the report. Figure 16–3 illustrates the format for a table of tables and figures.

  If a separate table for tables and figures is prepared, the format is almost identical to the one illustrated. The title is changed to "Table of Tables" or "Table of Figures." The margin headings "Tables" and "Figures" shown in Figure 16-3 are omitted.

---

**TABLE OF TABLES AND FIGURES**

**Tables**

| | Page |
|---|---|
| I. Comparative Costs of Construction at Sites A, B, and C | 6 |
| II. Comparative Costs of Maintenance at Sites A, B, and C | 8 |
| III. Comparative Market Potential at Sites A, B, and C | 11 |

**Figures**

| | |
|---|---|
| 1. Drawing of the Proposed Building | 4 |
| 2. Graph of Population Trends at Sites A, B, and C | 10 |

iv

---

**FIGURE 16-3**  An example of a table of tables and figures

The format for a table of tables or a table of figures is explained in the following guidelines.

- The title for the table is capitalized and centered horizontally 2 inches below the top of the page. Three spaces separate the title from the list.

- The table or figure numbers are listed on the left-hand margin. The numbers are followed by the table or figure title. The page on which the table or figure is presented in the report is listed on the right-hand margin.

- The side margins are the same as those used for reports. The list does not extend below one inch from the bottom of the page. An appropriate small Roman numeral to identify the page is horizontally centered in the bottom margin.

- A double space separates one listing from another. If the title of a table or a figure extends beyond one line, the title is single-spaced. The first letter of all important words in each title is capitalized.

# ABSTRACTS

An abstract to a formal report is a short but complete version of the report. In this respect, an abstract seems to duplicate a letter of transmittal. However, a letter of transmittal is usually conversational in tone and stresses ideas that are important to the reader. An abstract, like a report, is an objective, brief statement of basic information. If a long letter of transmittal and an abstract for a specific report seem to duplicate each other, the shortened letter of transmittal may accompany an abstract. A statement in the letter refers the reader to the abstract.

## Uses for an Abstract

Rules specifying when an abstract should be used have not been clarified. The writer's judgment; company policy; the length, content, and complexity of the report; and the reason for preparing the report help determine the need for an abstract. Although an abstract is occasionally omitted because the writer feels that a condensed version of the information may be misleading, most readers appreciate a well-written abstract for two reasons:

1. It gives an overall view of the text that makes reading the complete report easier.
2. It permits an executive to get general information from many reports without studying details of each report.

## Preparation of the Abstract

An abstract, as stated previously, is brief — only 5 to 10 percent as long as the finished report. The most meaningful abstract includes the purpose of the report, the important information related to each major topic, and a condensation of conclusions and recommendations. Emphasis is often placed upon conclusions because conclusions are the results of a summarized text.

An informational abstract prepared from the analytical report in Figure 10-2 is illustrated in Figure 16-4. The illustration is not representative of all informational abstracts. Because reports differ in purpose, content, and amount of detail, informational abstracts also differ. However, the following guidelines for preparing abstracts should be observed.

- The abstract is written after the report is completed. As nearly as possible, the organization of the abstract duplicates that of the report.

- The abstract is written in standard English. Telescopic writing, abbreviations, and symbols are avoided.

---

**ABSTRACT**

Thirty-five-millimeter photographs of metal surfaces were taken through Microscope A and Microscope B to determine which microscope should be purchased for use in stereophotography. A magnification ratio of 30 to 1 is required. The analysis resulted in the following conclusions:

1. Both microscopes achieved the required 30-to-1 magnification ratio. However, when Microscope A was used, the ratio was obtained only through a two-step enlargement of the negative.

2. Pictures taken with each microscope showed good detail of metal surfaces and adequate depth of field.

3. Critical focusing with Microscope A cannot be guaranteed.

4. Pictures taken through Microscope B were sharper and less grainy than those through Microscope A.

5. Microscope B is superior to Microscope A but more expensive.

Microscope B should be purchased because it produces superior pictures and saves 50 percent of the photographer's time.

---

FIGURE 16-4  An example of an abstract for a formal report

- An abstract makes no reference to the report. For example, an abstract, in contrast to a letter of transmittal, does not include a statement similar to "The attached report is an analysis of 35-millimeter photography." A writer might think of an abstract as a brief report.

- The text of an abstract should make sense. Transitional words, numerical listings, and careful groupings of ideas can interrelate one major idea to another.

- An abstract should be limited to a single page.

- Rules for the format vary. The general rule is that an abstract should be attractively arranged on a single page. The word *abstract*, centered horizontally 2 inches from the top of the page, is written in capital letters. Three line spaces separate *abstract* from the text. Side and bottom margins are the same as those used for reports. Abstracts are usually double spaced. However, a one-page, single-spaced abstract with double spacing between paragraphs is preferable to a two-page, double-spaced abstract.

# GLOSSARIES

A glossary attached to a formal report is a list of definitions of technical terms. Unit 3, "Definition in Reports," stresses the importance of defining unfamiliar technical terms in the report introduction. However, a writer, after considering the reader and the number of possibly unfamiliar terms in the report, decides whether to use a glossary. If a glossary is used, it and the page on which it begins are mentioned in the report introduction.

The word *glossary*, centered horizontally 2 inches from the top of the page, is written in capital letters. Three spaces separate *glossary* from the first definition. Each definition is written as a single-spaced paragraph. The term being defined is written as the paragraph heading. Two line spaces separate the paragraphs. Formal definitions, discussed in Unit 3, are used in glossaries; however, when necessary, the formal definition may be followed by an additional but brief statement of explanation. See "Focusing Screen" in the following example.

**EXAMPLE**

**Aperture.** An aperture in photography is the opening of a lens through which light can be transmitted.

**Focusing Screen.** A focusing screen is a ground glass mounted in the viewfinder of a camera. It is used to align the camera with the object to be photographed and to check the sharpness of the image that the lens produces.

**Prism Changer.** A prism changer is a device used to change the angle of view in stereomicroscopes from one eyepiece to the other.

The side and bottom margins for the glossary are the same as those used in the report. The terms being defined are listed alphabetically.

# APPENDIXES

An appendix refers to supplementary material usually attached at the end of a piece of writing. An appendix to a report contains only reference materials that are more distracting than helpful if used as part of the text. These reference materials include information such as specifications, statistical data, parts lists, conversion tables, and similar types of detailed information. Frequently, a large foldout such as a blueprint, a map, or a diagram is placed in an appendix where it can be opened and studied while the report is being read. Other graphic aids, relating to specific report topics, are placed within the report.

Each different type of reference material should be placed in a separate appendix. Each appendix is identified by a capital letter beginning with A. Either in the report introduction or in an appropriate part of the discussion, the reader is told what kind of material each appendix contains.

Appendix formats vary because information placed in appendixes varies. If the information is printed, a dividing page that has *appendix* written in capital letters and centered horizontally separates the appendix from the report. If the appendix contains typewritten information, the format is the same as that used for the report.

# ORGANIZATION OF REPORT SUPPLEMENTS

A title page is used for all reports containing introductory supplements. The supplements follow the title page in the following order:

1. Letter of transmittal
2. Table of contents
3. Table of tables
4. Table of figures
5. Abstract

The small Roman numerals assigned to introductory supplement pages are horizontally centered 1 inch above the bottom of the page. Glossaries and appendixes follow the report. Glossary and appendix pages are identified by Arabic numerals that consecutively follow the page numbers used in the report. The numbers are placed in the upper right-hand margin of the page.

# SUMMARY

1. Supplements, which contain special information about report content, are included with most reports. The six possible attachments are listed on page 183. The order for arranging introductory supplements is shown on page 193. Glossaries and appendixes are placed at the end of the report.

2. Nearly all organizations require that a report writer prepare a letter of transmittal, a table of contents, a table of tables or figures or both, and an abstract. This unit explains and illustrates each of these attachments.

3. The information in a letter of transmittal should not duplicate that in an abstract. When an abstract is submitted, and one usually is, the letter of transmittal is short, referring the reader to the abstract for summarized data, or has a conversational tone, emphasizing ideas that interest the reader.

4. Abstracts are important to executives who receive numerous messages. Unlike letters of transmittal, they never refer to the report; they are objective summaries of report content. Unless the writer can justify a change, they follow the organization of information used in the report.

5. Glossaries are used only for an extensive list that defines technical terms. Appendixes contain various reference materials that, if used within the text, would be distracting.

6. The same logical organization, clarity, conciseness, accuracy, completeness, consideration for the reader, and judgment used in preparing reports is used in preparing supplements.

UNIT **16**

# SUGGESTED ACTIVITIES

A. Using the information from a report written previously, prepare a letter of transmittal. You may assume that the report was requested or that you initiated it. In the opening paragraph, identify the report and the purpose to be accomplished.

Determine what kind of information will interest the recipient, and discuss the information in one or two paragraphs.

Assume that another person assisted in preparing the report or submitted information for the report. Write a paragraph giving that person recognition for his or her work.

Decide whether an additional closing paragraph is needed in the letter. If one is needed, be creative in expressing your ideas; avoid stereotyped expressions originated by others. If a closing paragraph is not needed, complete your letter with a complimentary close.

Duplicate the format used in Figure 16–1 and type the letter. Sign the letter in black or dark-blue ink.

B. Prepare an abstract for one of the reports you have written previously. Make the abstract a brief, objective, and coherent sketch of the major ideas expressed in the report. Using the format in Figure 16–4 as a guide, prepare the final draft of the abstract. The backing sheet discussed in Unit 7 has the correct margins for abstracts, tables of contents, tables of tables, tables of figures, and glossaries. The horizontal line 2 inches from the top of the backing sheet is used for all supplement titles.

C. Assume that all attachments discussed in this unit are needed for a report you have written. Using this information, prepare a table of contents.

D. Prepare a table of tables or a table of figures that contains the numbers, titles, and pages for four or more graphic aids.

E. Prepare a glossary containing the definitions of five or more technical terms that could be used in a single report.

# 17

# Research and Documentation

After reading this unit, you should be able to do the following tasks:

- Describe the uses for researched data.
- List and describe the methods of researching data.
- Obtain permissions and prepare credit lines for information received from other sources.
- Prepare parenthetical references.
- Prepare bibliographies.

## OVERVIEW

*Research* and *documentation* are general terms that have a variety of meanings. The discussion of these two terms in this unit will be limited to uses for researched data, methods of researching data, permission to use and credit for the use of researched data, references, footnotes, reference sheets, and bibliographies. The information in this unit relates only to that needed for technical writing.

# USES FOR RESEARCHED DATA

Research in technical writing refers to any careful and critical search for information used primarily for one of two purposes:

1. To substantiate (verify or prove) statements made in a letter or report.
2. To form a basis for a decision.

Researched data is often used in analytical reports, where verifications are needed to justify recommendations. Some analytical reports are based almost entirely upon research. For example, a social worker may research improved methods for testing job applicants. One organization planning a training program researches training programs conducted by other organizations. An employee of one industrial firm researches several competitive products to determine market potential for a proposed product. These reports are analytical because the writer is expected to use the information to formulate conclusions and to make recommendations.

# METHODS OF RESEARCHING DATA

Researched data for technical communications is usually obtained from one or more of three different sources:

1. Conversation with experts in a particular field
2. Letters requesting information from persons or firms qualified as experts in a particular field
3. Research in general and technical libraries

## Personal Interviews

A writer planning to obtain information for a report from another person makes an appointment. While arranging the time, the writer also explains the purpose for the interview and suggests the approximate amount of time needed for discussion. Before the interview, the writer lists the topics to be discussed and the questions to be asked.

During the interview, the writer takes written or taped notes, carefully recording exact figures, dates, and other specific data. A tape recorder is used only with an interviewee's permission. After completing the report, the writer usually asks the person who gave information to verify the writer's interpretation.

## Research Through Letters

A writer researching information through a letter uses a letter of request as explained in Unit 20. Before writing the letter, he or she outlines the topics and determines the type of information needed. Then specific questions relating to the information are formulated. Specific questions are better than general ones because they clarify the request and simplify the recipient's responses. Some requests contain a questionnaire to be filled in by the recipient. Questionnaires and requests attached to them must be skillfully prepared. Otherwise, many busy executives will not respond. Requests are accompanied by a self-addressed, stamped envelope.

## Library Research

Most students, through library and English courses, have studied procedures for researching materials in libraries. They understand how to use card catalogs, periodical guides, and other reference materials. Therefore, the discussion of library research in this unit is restricted to a suggestion for recording and organizing information so that it can be incorporated logically into a report.

Many textbooks concerning library research explain methods for organizing information. Any method that the researcher has already learned to use successfully is acceptable. A researcher who has no favorite method will find the use of the cards shown in Figure 17–1 effective for obtaining or organizing library information. Before beginning any library research, however, the writer makes a list of the specific data he or she needs.

The following guidelines were used to prepare the cards shown in Figure 17–1.

- Paper or index cards measuring approximately 3 by 5 inches are used. This size is easy to handle and is usually large enough to contain all information obtained from a single source.

- An Arabic numeral is assigned to each source researched. This number is placed in the upper right-hand corner of the publication card (the card containing the name of the source and information about the source).

- Beginning on the left-hand margin of the publication card, each of the following items is listed on a new line separated by double spacing from the line above it: the author; the name of the publication; the city (and state if the state is given on the source's title page) of the publication, the publishing company, the date of publication; the library in which the publication was found and the card catalog number of the publication. The last two items are placed on the same line.

- If an encyclopedia, government publication, or other document has no author, the card lists the name of the publication and other information needed for a footnote, a reference sheet, or a bibliography (see pages 201–208).

**PUBLICATION CARD**

                                                                        1

Author's name

Title of the publication (underlined)

City:     Name of publisher, Date of publication

Name of        PR (Card catalog number)
library        925
               H8
_____

**INFORMATION CARD**

p.83                                                                    1a

    In a shunt motor, the field windings are connected directly across the power source.  Therefore, their resistance must be high to keep the value of the field current low.

    Caution:  "If the field circuit of a shunt motor is broken, so that the flux approaches zero, the motor speed will increase to a dangerously high value."

    [Caution parallels warning found in *Motors*.]

**FIGURE 17-1** Examples of publication and information cards for library research

■ Notes taken from the researched text are written on separate cards called information cards. The page number of the source is placed in the upper left-hand margin of this information card. The number identifying the source, taken from the right-hand margin of the publication card, is placed in the upper right-hand margin of the information card. A small alphabetical letter follows the number. A separate information card is prepared for each new thought or quotation. The right-hand number on the information card identifies the publication; the small letter tells how many cards of information have been used from a single publication. For example, if additional information is taken, either from page 83 or another page, a new card is used. This card is numbered 1 b.

■ Information to be quoted directly from the publication is placed in quotation marks on the information card. Information converted to the researcher's words is written without quotation marks.

- Frequently, a person engaged in library research will formulate original ideas from the study of printed text. These thoughts are often forgotten unless they are recorded. Recording these ideas in brackets on the information card reminds the report writer of them later when he or she is preparing the report.

- When all research is completed, the cards are arranged so that all information from one publication can be placed behind the appropriate publication card.

The research method illustrated in Figure 17–1 eliminates duplicate publication cards, reduces the number of cards needed for research, and arranges information systematically by publications. It also prepares cards that may be used as references for footnotes and bibliographies. A quick review of the information tells the researcher whether there are duplications of information and whether the information obtained is adequate to accomplish the established purpose. This card arrangement, however, is not intended as a technique for organizing a report. After major topics of the report are determined, the information cards must be rearranged according to subject matter. For example, five different sources may contain information about shunt motors; yet shunt motors is one major topic.

# PERMISSIONS AND CREDITS

*Documentation* refers to the use of authoritative sources of information to substantiate statements made in a technical paper or to form the basis of a paper. Receiving permission to use other people's words or ideas and giving credit to those people for their assistance in the report preparation are part of the ethical standards required in technical writing.

A writer who obtains information for a report through a personal interview gets permission from the person interviewed to use the information. Usually, oral approval is adequate. Credit for the information is given in the letter of transmittal and in "Sources of Data" in the introduction to the report. A person assisting in the preparation of a company manual is given credit in the preface of the manual. If many people assist, an acknowledgment page may be attached to the end of the report.

A writer obtaining information through letters includes a statement that tells how the information will be used. People who respond to such letters indicate by their responses that they allow this use. Credit for the information is given in the letter of transmittal and in "Sources of Data." If only a few people respond to the letter and express interest in the report, a copy of the finished report is sometimes forwarded to them.

Fair-use practice permits report writers to quote directly or paraphrase ideas from copyrighted sources without obtaining written permission. However, this practice must be respected. The originating author must never be quoted out of context. The original meaning of the text should not be changed. A complete work or a portion that diminishes the value of a complete work should not be used without receiving written permission. Also, the copyrighted material must be documented in parenthetical references or footnotes.

# PARENTHETICAL REFERENCES

Many organizations have found that using footnotes to document quoted or paraphrased published data is time-consuming for the report writer and stenographer and also distracting to a reader. They, therefore, have adopted a documentation procedure called parenthetical referencing. The following procedure is used: A reference sheet is prepared. This sheet is identical to a bibliography (see Figure 17–2 shown later in this unit), except that each entry is numbered and the title is changed to "Works" (or "Literature") "Cited." See the following example.

**EXAMPLE**

**WORKS CITED**

**Books**

1. King, Richard Allen. <u>The IBM PC–DOS Handbook</u>. Berkeley, Paris, Dusseldorf: Sybex, 1983.
2. Lien, David A., and Gary Williams. <u>MS–DOS: the Basics</u>. San Diego: Compusoft Publishing, 1983.

**Periodicals**

3. Grevstad, Eric. "MS–DOS Scripsit: Honorable Mention, but No Prize," <u>80 Micro</u>. 75 (April 1986), 27–28.

The references are listed chronologically as they first appear in the report text. When used, the list replaces a bibliography. From this reference sheet, the number that identifies the source of information used in the report is selected. In parentheses following the quoted or paraphrased material, the number is written and followed by a colon. A number identifying the page on which the information was found follows the colon. Here is an example.

**EXAMPLE**

"You can have different headers and footers for odd and even pages, and change them at will" (3:27).

Some organizations, especially in the humanities, use the author's last name and the page: (Grevstad:27). Then, the works cited are not numbered. However, if an author is listed twice, some method such as the date of the publication referenced must identify the specific reference.

# FOOTNOTES

Footnotes, like references, are used to tell the source of a quotation or idea used within a report; however, footnotes may also be used to present supplemental information. Using a footnote to refer a reader to a definition or supplementary information is usually more distracting than helpful. Footnotes in formal reports, then, are used primarily to document quoted or paraphrased information. The discussion of footnotes in this unit is limited to the basic format for footnotes and to examples of footnoting the kinds of publication most frequently used in reports. Many manuals of style that give detailed explanations of footnotes and bibliographies have been published. One of these manuals should be available to a technician whose reports are dependent upon extensive library research. For suggested references, see the Suggested Supplementary Readings for this section.

## Format for Footnotes

Footnotes are identified by superscript (written above and to the right or left) numbers. However, either of two numbering techniques may be used: Footnotes are numbered consecutively through the report, or each new page begins with the Arabic numeral 1. If fewer than ten footnotes are used in the entire report, consecutive numbers are preferred. Consistency is more important than technique. In reports containing sections or chapters, numbers are usually consecutive throughout one section or chapter. Arabic numeral 1 begins the footnotes for each new section or chapter.

Each quoted or paraphrased statement, partial statement, or paragraph written in the text is followed by a superscript number. A corresponding superscript number precedes the footnote relating to the information. The following guidelines are used to arrange footnotes at the bottom of the page containing the referenced material. (Some writers place all footnotes on a single page at the end of a section, chapter, or report. However, this arrangement has not received general approval.)

- The underliner key on the typewriter is used to type a solid line extending 2 inches from the left-hand margin. This line is placed one space below the text.

- Footnotes begin two spaces below the solid line. The backing sheet discussed in Unit 7 may be used to determine the amount of space needed to write footnotes. However, the solid dividing line in Figure 7–1 is adjusted upward or downward according to the number of footnotes needed. A 1-inch bottom margin must be left below the final footnote. A footnote must be placed on the same page as the superscript number identifying the referenced material. For example, material may begin on page 8 and continue to page 9. The footnote for this material is placed on page 9.

- Information in each footnote is single-spaced. The first line is indented five spaces from the left-hand margin. A superscript number corresponding to the superscript following the quoted or paraphrased information used in the report precedes the footnote. A double space separates one footnote from another.

- Internal punctuation for footnotes is illustrated in the following example.

| | |
|---|---|
| **EXAMPLE** | [1]Ted Schwarz, <u>**Careers in Photography**</u> **(Chicago: Contemporary Books, Inc., 1981), p. 81.** |

The name of the author is written as it appears on the title page of the textbook. (If an author's full name has been used in the report to introduce the reference, the author's name may be omitted in the footnote.) Commas separate one major item in a footnote from another. The city in which the book is published, the publisher, and the date of publication are enclosed in parentheses. The name of the city is followed by a colon. A comma separates the publisher's name from the date. A comma follows the closing parenthesis. The page number is designated by a lowercase $p$ if the information was obtained from a single page or by $pp$ if the information came from more than one page. Both abbreviations are followed by a period. A period is also placed at the end of the footnote.

- If a state is included with the city on the title page, the state, written as it appears on the title page (abbreviated or spelled out), follows the name of the city. The name of the state is separated from the city by a comma; the colon follows the name of the state. See the next example.

| | |
|---|---|
| **EXAMPLE** | **(Englewood Cliffs, New Jersey: Prentice-Hall, 1983)** |

- Complete works — textbooks, journals, magazines, or reports — regardless of length, are underlined. Parts of a complete work — sections, chapters, or articles — are placed in quotation marks. The first letter of all important words in all titles is capitalized.

## Examples of Footnotes

The following footnotes illustrate the kinds of references usually documented in reports and company manuals. The first four examples refer to textbooks; one example refers to a magazine article; the final example refers to unpublished papers.

A footnote for a textbook written by one author is styled as shown in the following example.

| EXAMPLE | ¹Robert E. Swindle, The Concise Business Correspondence Style Book (Englewood Cliffs, New Jersey: Prentice-Hall, Inc., 1983), p. 141. |

A footnote for a textbook written by two or three authors is styled as shown in the following example.

| EXAMPLE | ¹Walter E. Oliu, Charles T. Brusaw, and Gerald J. Alred, Writing That Works (New York: St. Martin's Press, 1984), pp. 266–272. |

If more than three authors write a textbook, the author entry is written as Carl J. Jason and others.

When a textbook is prepared by a staff of writers and no editor is named on the title page, the footnote is written as shown in the following example.

| EXAMPLE | ²The Chicago Manual of Style (Chicago: The University of Chicago Press, 1982), p. 236. |

A footnote for a textbook that has no author but has an editor listed on the title page is written as shown in the following example.

| EXAMPLE | ³James Winters (ed.), Title of Textbook (City: Publisher, date), p. 19. |

Footnotes of articles found in a magazine, a journal, or a similar type of publication are written in one of two ways. When the author of an article in a journal or magazine is indicated, the footnote is written as shown next.

| EXAMPLE | ⁴Doug DeMaw, "Measuring Small-Value Capacitors," QST, LXX (Newington, Conn.: American Radio Relay League, September 1986), p. 31. |

As indicated in the example, the author, the name of the article, the name of the publication, the volume number, the publication information including the month and year, and the page used from the article are given. If an article shows no author, the footnote includes all information given in the foregoing example except the author's name.

Report writers frequently use reports, manuals, and dissertations prepared by professional people or experts in various fields of technology. Many of these documents are unpublished. The titles of unpublished material are placed in quotation marks. These documents are footnoted in the following way.

---

**EXAMPLE**

[5]**Author, "Title of the Report, Manual, or Dissertation" (College, company, or organization for whom the work was prepared, date), page reference.**

---

If the document is divided into sections or chapters that have been given a title, the section or chapter title, placed in quotation marks, precedes the name of the document.

## Shortened Footnote Forms

The term *Ibid.* is an abbreviation for the Latin term *ibidem*, which means "in the same place." When the same source is used for two or more consecutive footnotes, the term *Ibid.* replaces the previously stated information about the source. Notice that *Ibid.* is always capitalized, underlined, and followed by a period. When the information is taken from the same source but from a different page, a comma and page reference follow the term. When the information is taken from the same source and the same page, the term is used without a page reference. See the example that follows.

When a footnote for a different source comes between two references for the same source, the previously listed author's name is used instead of *Ibid.* The author's name is followed by a page reference. If the author is unknown, the title of the work is used instead. (The use of op. cit., the Latin term that means "in the work cited," is another way to footnote a previously written-out source. In recent years, however, the first method has been preferred.)

---

**EXAMPLE**

[1]**Richard Allen King, The IBM PC–DOS Handbook (Berkeley: Sybex, 1985), p. 295.**

[2]**Ibid., p. 200.**

[3]**Ibid.**

[4]**David A Lien and Gary Williams, MS-DOS: The Basics (San Diego: Compusoft Publishing, 1985), p. 99.**

[5]**King, p. 202.**

---

# BIBLIOGRAPHIES

A bibliography is a compilation of all materials used in a report and listed in the parenthetical references or footnotes. It does not include sources that were consulted but not used to write the report. The entries in a bibliography attached to a report refer to entire works in which the articles or pages used may be found, not the specific articles or pages alone. The bibliography is placed at the end of the report. No rules have been established for the location of the bibliography in relation to a glossary or an appendix. However, a report rarely, if ever, contains all three supplements. If one does, the bibliography could precede the other two supplements.

Organization of bibliographies varies. In most reports, bibliographies are divided into major categories such as textbooks; journals, magazines, and similar references; and unpublished text. Each list is placed under a separate heading. The references under each heading are arranged alphabetically by the author's last name. Therefore, the author's name, in contrast to the form used in footnotes, is written with the last name placed first. If the reference gives no author, the name of the text is used to alphabetically place the book. See the fourth listing in Figure 17–2. Figure 17–2 illustrates a bibliography used in a formal report. However, when writing reports that contain bibliographies, technicians should consult manuals of style for detailed information.

The format for a bibliography is arranged according to the following guidelines.

- *Bibliography*, horizontally centered and written entirely in capital letters, is placed 2 inches below the top of the page.

- If the bibliography is subdivided into books, periodicals, and unpublished papers, an appropriate subheading is horizontally centered and placed three spaces below the preceding text. A double space is used below the subheading. The first letter and all important words of the subheading are capitalized.

- The first entry begins flush with the left-hand margin of the page. (The backing sheet may be used as a guide.) All written matter relating to one entry is single-spaced. The second and each succeeding line of a single entry is indented five spaces from the left-hand margin.

- A double space separates one entry from another.

- A book entry is punctuated as shown in the next example.

**EXAMPLE**

Author's Last Name, First Name or Initial. <u>Title of the Book</u>. Place of Publication: Publisher, Copyright Year.

- A periodical entry is punctuated as shown next.

BIBLIOGRAPHY

Books

Gay, Peter. <u>Education of the Senses</u>. New York: Oxford University Press, Inc., 1984.

Gould, Stephen J. <u>The Flamingo's Smile: Reflections in Natural History</u>. New York: W. W. Norton and Company, Inc., 1985.

Sagan, Carl, and Ann Druyan. <u>Comet</u>. New York: Random House, Inc., 1985.

<u>The Galaxy and the Solar System</u>. Smoluchowski, R., J. N. Bahcall, and M. S. Matthews, editors. Tucson: University of Arizona Press, 1985.

Government Publications

Morrison, David, et al. <u>Planetary Exploration through the Year 2000: A Core Program</u>. Washington, D.C.: U.S. Government Printing Office, 1983.

<u>The Cosmic History of the Biogenic Elements and Compounds</u>. Wood, John, and Sherwood Chang, editors. NASA Special Publication No. 476, 1985.

Newspaper

Wilford, John Noble. "U.S. and Soviets Cooperating on Collection of Comet Dust." <u>New York Times</u>, December 21, 1984.

Periodicals

Angier, Natalie (reported by William Dowell/Paris, Jon D. Hull/Los Angeles, and Thomas McCarroll/New York). "Greeting Halley's Comet." <u>Time</u>, 123, No. 24 (December 16, 1985), 60-70.

Kerr, Richard A. "Mars Is Getting Wetter and Wetter." <u>Science</u>, 255, No. 3 (September 1986), 92-99.

FIGURE 17-2   An example of a bibliography for a formal report

---

**EXAMPLE**

Author's Last Name, First Name or Initial. "Title of the Article." <u>Title of the Periodical</u>, Volume Number (complete date of issue), page(s).

- For parenthetical references, as noted on page 201, Arabic numerals precede each listing, and the entries are arranged chronologically according to first use, regardless of the different kinds of published sources listed. The title is usually changed to "Works" ("Literature") "Cited" instead of "Bibliography."

For additional examples of bibliographies, look at the Suggested Supplementary Readings at the end of each section of this text.

# SUMMARY

1. Research is any careful and critical search for data used primarily to verify or prove statements made in a written communication and to form a basis for a decision. Researched data is obtained through personal interviews, letters that usually include a questionnaire, and library sources.

2. A card system or similar note-taking procedure, explained and illustrated in this unit, is essential to effective library research. Notes or a tape recorder are also suggested for interviews if the person being interviewed approves.

3. Any information quoted or paraphrased from published sources must be documented. Parenthetical references or footnotes, discussed and illustrated in this unit, are used for this purpose. Company or organization policy usually determines which is used.

4. Bibliographies, often referred to as "Works" ("Literature") "Cited" when parenthetical references are used, list all published sources from which information in the report is taken. Published sources researched but not used are not listed.

5. Bibliographies for parenthetical numbered references must include Arabic numerals placed chronologically before each entry.

6. Style manuals vary in suggesting methods for placing footnotes and bibliographies in reports. Individual companies and organizations usually have policies the report writer adopts.

UNIT **17**

# SUGGESTED ACTIVITIES

A. Select a subject that requires library research. Make an outline for the topic so that the information requiring research is apparent. Then list the topics to be researched. Using the card technique illustrated in Figure 17-1 or a research technique of your own, research at least five different authors. Find at least two quotations or paraphrases from each author. Some of the materials may be taken from a text that does not indicate an author.

B.  Write a one- or two-page paper in which the researched information can be used. Superscript the information used in the text and prepare appropriate footnotes. Paraphrased material is incorporated into the text. No quotation marks are used. Quoted material is written according to the rules for quoted material explained in Unit 7.

C.  Prepare a bibliography of researched materials that could be used in a single report — all materials will relate to one general subject.

D.  Using the same material as that for B and C, change footnotes to parenthetical references and the bibliography to a "Works Cited" sheet.

# SUGGESTED SUPPLEMENTARY READING

Damerst, William A. *Clear Technical Reports*. New York: Harcourt Brace Jovanovich, Inc., 1972. In the discussion of abstracts, pages 121 to 124, ten specific steps for preparing an informative abstract are offered. Also, on pages 71 to 87, research and note-taking techniques are discussed.

Houp, Kenneth W., and Thomas E. Pearsall. *Reporting Technical Information*. New York: Macmillan Publishing Company, Inc., 1984. Report supplements are discussed on pages 201 to 214 (Appendixes, page 227). Use of the library is explained and illustrated in Chapter 4. Methods for gathering and checking information are discussed in Chapter 5. Documentation is discussed on pages 232 to 241.

Lannon, John M. *Technical Writing*. Boston: Little, Brown and Company, 1985. The author discusses the following: report supplements, pages 240 to 253; library research, pages 282 to 294; interviews and questionnaires, pages 294 to 305.

Markel, Michael H. *Technical Writing: Situations and Strategies*. New York: St. Martin's Press, Inc., 1984. On pages 28 to 52, the author discusses library research and lists various reference sources available in the library. He also discusses interviews, questionnaires, and note taking. On pages 233 to 260, he discusses report supplements.

Roundy, Nancy, with David Mair. *Strategies for Technical Communication*. Boston: Little, Brown and Company, 1985. On pages 366 to 388, report supplements are discussed and illustrated. On pages 42 to 69, the author details procedures for library research.

Sherman, Theodore A., and Simon S. Johnson. Englewood Cliffs, N.J: Prentice-Hall, Inc., 1983. Report supplements are discussed on pages 233 to 234, and 245 to 250.

The following four manuals are authoritative sources of information concerning footnotes and bibliographies. (Any recent edition may be used.)

Campbell, William Giles, and Steven Vaughn Ballou. *Form and Style: Theses, Reports, Term Papers*. Boston: Houghton Mifflin Company.

Keithley, Erwin M. *A Manual of Style for the Preparation of Papers and Reports*. Cincinnati, Ohio: South-Western Publishing Co.

Turabian, Kate L. *A Manual for Writers of Term Papers, Theses, and Dissertations*. Chicago: The University of Chicago Press.

The University of Chicago Press. *The Chicago Manual of Style*. Chicago: The University of Chicago Press.

# Writing Letters

# 18

# Techniques for Effective Letters

## OBJECTIVES

After reading this unit, you should be able to do the following tasks:

- Discuss four objectives of business letters.
- Describe the "you" attitude.
- List and describe the six traits that contribute to a positive tone.
- Discuss the aids to effective use of language in letter writing.
- Outline the features of an attractive letter format.

## OVERVIEW

Technical students frequently assume that company executives write most business letters. This assumption may be true. However, executives write letters concerning an entire organization's operations. Technicians can expect to write inquiries, requests, and acknowledgments concerning assistance, information, and merchandise related to their work. They also write letters refusing assistance, information, and merchandise. Letters are exchanged among employees of one company or are written from one company to another. Even technicians who make oral requests to their supervisors are often asked to "Put it in writing" so that records of the requests may be filed. Therefore, the ability to write an effective letter and to write it without frustration or wasted time is an asset to any employee in any organization.

This unit gives some general techniques that help improve and simplify all letter writing. Techniques used in specific types of letters are discussed in Units 19 to 23.

# GENERAL OBJECTIVES

Technicians who understand some basic objectives to be accomplished in all business letters can more easily understand their role as a correspondent. The following list explains four of these objectives.

1. Every business letter is a sales letter. It is written to sell a product, an idea, a proposal, a request for assistance or advice, or goodwill between companies or individuals. If a letter of refusal is written, the letter sells the idea that the refusal is reasonable and logical.
2. The real test of a successful letter is whether it accomplishes the purpose for which it is written.
3. A letter is a written substitute for oral communication. The rules of courtesy and mutual respect that apply to personal relationships apply to business letters.
4. A letter is a written record of business transactions that reflect the writer's attitude toward, understanding of, and interest in the company or organization he or she represents.

Writers achieve the objectives they strive for in their letters by developing a favorable attitude toward letter writing; by using effective, grammatically correct sentences; and by making the format of the letter attractive.

# THE WRITER'S ATTITUDE

In the same way that a speaker's attitude is reflected in manner and voice, a correspondent's attitude is reflected in the method used to express an idea in a letter. Every correspondent wants to be the kind of person that a reader can like, can work with congenially, and can trust. A writer accomplishes these goals by substituting a "you" attitude for an "I" attitude and by making the tone of the letter positive and cooperative.

## The "You" Attitude

A writer who is open-minded enough to consider the reader will have little trouble incorporating a "you" attitude into letters. The writer will know that a reader becomes interested in any communication that reflects an appreciation for the reader's time, talent, and effort. The difference between the "you" and the "I" attitude is illustrated in the following two versions of the same idea.

> **EXAMPLE**
>
> **I would appreciate your cooperation in answering some questions related to a report I am preparing concerning the use of polymers in industry. ("I" attitude)**
>
> **Do you have a few minutes to help me verify some information needed for a report concerning the use of polymers in industry? ("You" attitude)**

A single example does not effectively illustrate the importance of developing consideration for the reader in all letters. Examples of complete letters used as illustrations in Units 19 to 21 add further emphasis to this important concept in business communications. The example here is adequate, however, to show that the "you" attitude is not accomplished simply by substituting *you* for *I*. It results from a genuine awareness that the reader's services, time, and goodwill are worth the attention given them by the writer.

## Tone

Tone is the manner of expression that transmits the feelings of the writer to the reader. If the writer is angry, the tone of the letter reflects anger. If the writer genuinely considers the reader, the tone of the letter reflects consideration. Many specific traits in a writer's attitude contribute to a positive, cooperative tone. The most important traits are courtesy, sincerity, cheerfulness, friendliness, confidence, and respect.

**Courtesy**    Courtesy, a trait easily recognized in letters, is the display of good manners. Courteous writers always consider the reader's point of view. Thus, they never write while angry. They understand that an angry letter creates one or two unfavorable reactions: The recipient responds in an angry tone; or the recipient refuses to answer, and goodwill, important in all business, is adversely affected.
Courteous writers also avoid abruptness, sarcasm, innuendoes, and accusations. They recognize the possibility of misunderstandings that can develop in written communication and thus graciously permit the reader to express a point of view. All statements reflect reason, tact, and thoughtfulness.

**Sincerity**    An attitude of sincerity strongly affects the tone of a letter. The wording is straightforward, specific, and convincing. Sincere writers give genuine praise but avoid flattery. They also avoid exaggerations, distortions, and vagueness. They write naturally, permitting their own personalities to become evident. Thus, their letters should never sound stilted, unduly familiar, apologetic, or uninteresting.

**Cheerfulness**    Cheerfulness is a trait every reader appreciates when it is detected in the tone of a letter. Cheerfulness and a sense of humor, appropriately incorporated into business communications, add warmth to the message and encourage cooperation from the recipient.

**Friendliness**  The tone of friendship results from a genuine desire to help the reader. A service attitude that indicates the writer understands the importance of cooperation in business is the basis for friendship.

**Confidence**  Confidence tells the reader that the writer is capable of making unbiased decisions. It permits the writer to express ideas convincingly without boasting and without lecturing to the reader or belittling the reader's point of view.

**Respect**  Respect is approval of and appreciation for the qualities possessed by another individual or company. Two people or companies cannot continue a compatible relationship unless mutual respect is reflected in the tone of the letters they exchange.

The tone of a letter is not controlled by a set of rules relating to the six traits just described. It results from the personality of the writer and the writer's ability to express his or her personality in business communications.

# EFFECTIVE USE OF LANGUAGE

In the units explaining formal reports, the objective, scientific, third-person style of writing is emphasized. This style is not recommended for business letters. Letters are conversations. Therefore, the language is conversational. The writer talks with the reader. In this respect, language in business letters is similar to the language in oral reports. The pronouns *I*, *we*, and *you* are used extensively. The active voice replaces the impersonal passive voice frequently used in formal reports.

Although sentences in letters are always grammatically correct, well organized, and concise, they are written in a relaxed, natural style that brings the reader into a personal relationship with the writer. The degree of informality in the tone of the letter, however, is determined by the degree of friendship and understanding that has developed between two correspondents. An occasional colloquialism may sometimes be more appropriate than a formal expression. Thus, the style of language used in a business letter is determined not only by rules but also by good judgment, a positive attitude, and an understanding of the reader.

Seven of the eight basic principles of technical writing discussed in Unit 1—understand the reader, know the purpose of the letter, know the subject matter, organize the material, use standard English, use correct format, and adopt ethical standards—apply to business letters as well as to reports. (The fifth principle, writing objectively, as explained previously, does not apply to letters.) The rules for correctness, conciseness, clarity, and completeness explained in Units 24 to 30 also apply to letters. Their importance as aids to writing effective letters is reemphasized in the following discussion of letter content, clarity, conciseness, completeness, and meaningless phrases and clichés.

## Letter Content

The organization and content of specific kinds of letters are discussed in Units 19 to 21. Three general considerations for the content of all letters are presented next.

1. A writer who wants a complete and specific response to a particular topic limits a letter to that topic. Introducing many important ideas into a single letter forces a division of the reader's interest and attention.
2. Every business letter, like every formal report, has a specific purpose to be accomplished. Before writing any letter, a writer should ask, "Is this letter necessary? Does the information it will contain overlap previous correspondence? What specific purpose is it designed to accomplish?"
3. Reading a completed letter should require no more than one and one-half minutes. In this length of time, a good letter writer has interested the reader, has convinced the reader that the subject of the letter merits consideration, and has stimulated the reader to respond.

## Clarity

Clarity in letters, like clarity in reports, begins with careful planning. Unfortunately, too many writers fail to recognize the value of preplanning. They prefer writing three or four drafts of a letter or risking the chance that a poorly written letter will bring a favorable reply. In preplanning, writers determine the purpose of their letters, visualize their readers, and organize their subjects. Making marginal notes on a letter being read, like making notes for any future reference, permits writers to answer letters specifically and completely.

All ideas in a letter are expressed in a straightforward style. Because a letter has no place for definitions, a good letter writer is skillful in choosing words that say precisely what is intended. Words are chosen to express an idea, not to impress the reader; therefore, specific words are usually better than general terms.

## Conciseness

An unrequested letter is, to some extent, an infringement upon a reader's time. Therefore, the writer is obligated to be concise. Conciseness is the elimination of unnecessary words and phrases, as explained in Unit 29 and further illustrated in "Clichés and Meaningless Expressions." Conciseness also refers to the omission of long, introductory paragraphs that talk around the purpose of the letter without conveying any important information. Conciseness, however, is not abruptness. Any words or sentences essential to completeness and clarity are included. Words needed to improve the tone of the letter are also used.

## Completeness

A letter containing incomplete information is expensive. If questions remain unanswered or necessary data is omitted, additional communication must supplement the original message. For this reason, writers should check their information for completeness before signing letters.

## Clichés and Meaningless Expressions

Clichés and meaningless expressions persist in letters even though several lists of overused words have been published in many sources throughout most of the twentieth century. Clichés are dull, unimaginative expressions that readers are inclined to bypass when they read the letter. Meaningless expressions are time-consuming and distracting. Unfortunately, most of these expressions are placed at the beginning or at the end of the letter — the parts reserved to arouse reader interest and stimulate response. Consider the following example, which is the opening statement of one business letter.

**EXAMPLE**

Please find enclosed a stock list showing all the sections at present held in our local office.

The two underlined parts are clichés. Ironically, the salutation of this letter is *Dear Bob*, an informal salutation that suggests that the correspondents communicate frequently. The clichés and the tone of formality in the sentence following the informal salutation seem contradictory.

The following example is quoted from an opening sentence in another business letter.

**EXAMPLE**

Pursuant to our telephone conversation, there is herein set forth some questions that are troubling our department relative to the recent offer of settlement with reference to the proposed construction of a right of way through certain properties of said client.

This sentence has many disadvantages. The underlined words are clichés. The verb *is* disagrees with its subject *questions*. The sentence contains twice as many words as the number recommended for sentences used in informative communications. The letter has a dictatorial and antagonistic tone resulting from an "I" attitude. The tone may be justified, but the important question is, "Will it help resolve a misunderstanding?"

Distinguishing between meaningless words and clichés in letters is often difficult. Many expressions may be classed as both. Examples of some clichés are given in Figures 29–1 and 30–3. Some additional ones are listed in Figure 18–1. The long but still incomplete list will help careful writers recognize and avoid clichés.

| | |
|---|---|
| **as of the date of** | **pursuant to your request** |
| **assuring you of our prompt** | **receipt is hereby acknowledged** |
| **  attention** | **respectfully requested** |
| **attached hereto** | **  (submitted)** |
| **at the earliest possible moment** | **said account (record, report)** |
| **at the present writing** | **subject to your approval** |
| **due to the fact that** | **this is to acknowledge** |
| **enclosed herewith (herewith** | **upon receipt of** |
| **  enclosed)** | **we are cognizant of the fact that** |
| **for your information** | **we are in receipt of** |
| **in the amount of** | **we hope this letter (plan, project,** |
| **in the near future** | **  report) meets with your approval** |
| **in this connection** | **we wish to acknowledge (state,** |
| **kindly favor us** | **  inform, advise)** |
| **permit me to say** | **your esteemed favor** |
| **please be advised** | **yours of recent date** |
| **please find enclosed** | |

**FIGURE 18-1**  Clichés to avoid in letter writing

# LETTER FORMAT

Anyone who has had the opportunity to read a business letter knows that a neat, attractively arranged format creates a favorable impression. In fact, the impression begins when the recipient first sees the envelope. Although format may not be the most important factor in letter writing, it does influence a reader's reaction to the content.

Students who have completed English courses or typewriting courses in high school have learned the components of business letters. Many explanations and examples of format are found in typewriting manuals, textbooks for letter-writing courses, English textbooks, and some dictionaries. Explanations and examples are also referenced in the Suggested Supplementary Readings at the end of this section. In addition, correct format is illustrated in the letter report in Unit 13 and in examples of letters in Units 19 to 21. As these illustrations show, more than one format for business letters is approved. Letter format in this unit is limited to guidelines that make a letter attractive and easy to read. These guidelines follow.

■ Company stationery is usually used for all letters written for the company. Company stationery contains the company's name and address. For this reason, most examples of letters in this book do not contain a place for such information. However, if the name and address is not printed on the stationery, the information is indented so that it is flush with the right margin, as indicated in the following example.

**EXAMPLE**

**Company Name**
**Street, City, State, Zip code**
**Month, Day, Year**

If company stationery is not used, technicians write their letters on the kind of paper specified for formal reports. See Unit 7. Envelopes preferably match the type of paper used for the letter. Two standard sizes of envelopes are available. One measures 9 1/2 inches by 4 1/8 inches; the other measures 6 1/2 inches by 3 5/8 inches. The larger envelope is preferred because it simplifies folding and unfolding the letter page.

■ Detailed rules for arranging margins for letters are explained in typewriting manuals and are taught in typewriting courses. The general rule for an attractive letter format specifies that the white margin around the letter serves as a frame. Preferably, a little more white space is left at the bottom of the page than at the top. Side margins are equal to each other. The right-hand margin is made as even as possible after rules for correct word division are observed.

■ If a letter extends beyond a single page, at least two lines of the letter must be placed on the second page and begin not less than 1 1/2 inches from the top of the page. The second page begins with the name of the recipient, the date, and the page number listed and single-spaced on the left-hand margin of the page. See the last example in Unit 13.

■ Each paragraph of a letter is single-spaced. A double space separates one paragraph from another. In the most frequently used business-letter format, paragraphs are not indented; they begin flush with the left-hand margin. However, if a writer prefers, each paragraph may be indented five spaces. Paragraphs in letters should be short but varied in length. The opening and closing paragraphs preferably do not extend beyond four lines. Internal paragraphs that extend beyond eight lines, should, if possible, be divided. In contrast to paragraphs in reports, a one-sentence paragraph in a letter is not unusual.

■ All letters are typewritten or printed by a word processor. Many aids are available for correcting errors. Therefore, marking over an error or failing to make clean erasures on typewritten text is inexcusable. A typewriter ribbon that makes dark, even, distinct print is used. The keys of the typewriter are clean so that the printed letters do not appear smeared. Word-processed text is printed only on letter-quality dot-matrix or daisy-wheel printers. Spelling errors on either typewritten or printed text must be corrected.

- The handwritten signature following the complimentary close should be legible and written in black or dark-blue ink.

- The envelope, like the letter, is addressed on the typewriter.

Some general letter-writing techniques necessary to the effectiveness of all letters were discussed in this unit. Specific techniques needed to organize and write each of three different classes of letters are covered in Units 19 to 21. The three classes are identified as positive letters, persuasive letters, and negative letters.

# SUMMARY

1. Technicians frequently write letters related to their work. They request information, services, or merchandise; express appreciation; or refuse a request. They are also likely to write letters of application.

2. Every business letter is a sales letter. Thus, any letter that fails to sell the idea, service, or merchandise it intended to sell wastes time, money, and energy.

3. Effective letters meet the standards for all communication advocated in seven of the eight basic principles of writing (Unit 1). Writing objectively is the exception. The tone of the letter is conversational, but any information is given completely and without bias or prejudice.

4. A positive attitude is essential for effective communication through letters. This attitude includes consideration for the reader reflected in a "you" attitude and a courteous, sincere, cheerful, friendly, confident, and respectful tone.

5. Overworked expressions including clichés (page 219) and needless words and phrases (Units 29 and 30) bore a reader, distracting him or her from the real purpose of the letter.

6. Attractive letter format is essential. Because it is the first thing a reader sees, it can predetermine the reader's attitude — positive or negative — toward the message.

UNIT **18**

# SUGGESTED ACTIVITIES

**A.** Rewrite the following sentences to eliminate unnecessary words or clichés. Be original. Express the sentences in your own words.

1. We are in receipt of your order and will ship the merchandise on February 10.
2. This is to acknowledge receipt of merchandise we received on February 24.
3. Due to the fact that the merchandise was damaged when it was taken from the loading dock, we want to bring to your attention that it is being returned.
4. We beg to inform you that according to our records your account in the amount of $45 is now past due.
5. We regret to advise you that the merchandise you ordered cannot be shipped for thirty days. We deem it advisable that you reorder at that time if you still want the merchandise.
6. This report is respectfully submitted herewith, and I sincerely hope that it meets with your approval.
7. Enclosed please find a copy of our latest catalog. I hope you avail yourself of the opportunity to study its contents.
8. I am awaiting your future orders. You may be assured that they will receive our prompt attention.
9. We received your check this morning and have duly credited the $30 to your account.
10. Please feel free to write again at your earliest convenience.

**B.** Revise the following sentences to improve the tone.

1. Send us immediately the following items:
2. As you will note from the enclosed price list, available sizes of cones will range from the 1 3/4-inch by 2-inch cone on up to and including the 9 7/8-inch by 3-inch cone.
3. We know that the performance of this grinder will thrill you.
4. Our primary concern is to maintain the highest-quality standards in packaging as well as product — and perhaps you can assist us in this endeavor.
5. Don't let another day pass without giving us an opportunity to tell you about our fine product.
6. I wish to congratulate you on your recent promotion.
7. You can understand, of course, that this faulty merchandise is costing us time and money.
8. It was a real shock to me to find your name among our delinquent accounts.

C. Assume that the letter shown in Figure 18-2 was written to a person who had complained about a snap-on cap that was difficult to remove from a bottle. Read the letter; then follow the instructions given next.

1. Find a statement that belittles the reader's judgment.

2. Find at least one statement indicating an "I" attitude.

3. Is the writer courteous? Justify your answer.

4. Find three or more examples of unnecessary words or clichés.

5. In a paragraph or two, explain what your reaction would be if you, as a customer, received this letter. Use statements in the letter to justify your reactions.

---

Dear Mr. Blank:

We have your letter of January 20th regarding the difficulty you had with the plastic applicator bottle.

The snap-on cap for the bottle should snap easily and securely into place once the top of the nozzle has been snipped off as shown in the instructions.  While we realize that many people have their own preferences for a particular type or style of applicator bottle, we are sorry that you found ours inconvenient.  This item was included in the kits so that the customer need not purchase one separately.

If the applicator bottle in your kit was unusually difficult, we would appreciate it very much if you would return it to the attention of the undersigned so that we could refer it to the proper department for examination.

If you return the applicator bottle, we'll send you another.  We feel certain that this bottle will perform to your complete satisfaction.

Thank you for writing and offering us this opportunity to be of service to you.  In addition to the new applicator bottle, we are also sending you one of our other fine products to try.  We hope you will enjoy it.

                                        Sincerely,

---

FIGURE 18-2  An example of a response to a customer complaint

# The Positive Letter

After reading this unit, you should be able to do the following tasks:

- Organize a positive letter.
- Write a letter of inquiry or request.
- Write a letter of acknowledgment.
- Write a letter of good news or goodwill.

## OVERVIEW

The positive letter is an order, an inquiry, an acknowledgment, or a similar type of day-to-day letter that requests or acknowledges assistance, information, or merchandise. It also includes good-news letters expressing approval or acceptance of a plan or product and goodwill messages expressing appreciation or congratulations. These letters are classed as positive because they rarely involve controversies. Most of them are concerned with subjects that are mutually beneficial to the sender and the receiver. For these two reasons, positive letters are easily prepared and seldom fail to bring a needed response. For these same two reasons, positive letters are often carelessly prepared. The writer is inclined to become abrupt, demanding, and inconsiderate. Clichés that are recognized and avoided in letters requiring more thoughtful planning are too frequently relied upon as techniques for completing a positive letter quickly.

To help writers avoid careless preparation of positive letters and to help create goodwill through these letters, this unit explains and illustrates techniques for writing them effectively.

# ORGANIZING THE POSITIVE LETTER

Occasionally, a positive letter is extremely brief. Only one or two statements are needed to complete the entire message. An example of this kind of message is a request for a catalog, a parts list, or a diagram of equipment recently purchased from a company. Many companies prepare these items and expect customers to request them. Therefore, the message may simply state, "Please send me your new fall catalog of power tools." If the writer is uncertain whether the company expects payment for the requested item, cost is mentioned in the message. The writer either encloses the amount or asks that the company submit a bill.

A similar brief message may simply say, "Thank you." A short message of this kind, personally signed by everyone who heard a speech, studied a subject, or received other assistance is appreciated by some recipients. Others may prefer a longer message, as explained in "Writing Goodwill Letters."

Usually, a positive letter is divided into three parts:

—The first part introduces the subject.
—The second part explains the subject.
—The third part stimulates the recipient to action, expresses a promise to act, or just adds a pleasant closing note to the letter.

Each of the parts is illustrated in the examples explained in "Writing Positive Letters."

# WRITING POSITIVE LETTERS

A positive letter begins immediately with the introduction of the subject to be discussed. The information is written concisely and specifically. This section illustrates and explains how the three parts of positive letters are organized and occasionally modified to write inquiries, acknowledgments, good-news letters, and goodwill letters.

## Writing an Inquiry or Request

The distinction between a letter of inquiry and a direct request is so slight that both are discussed as letters of inquiry in this unit. One simple inquiry for a catalog was illustrated previously. Most inquiries, however, consist of three parts. The letter begins with the reason for the inquiry and then specifies the subject about which the inquiry is made, as indicated in the following example.

**EXAMPLE**    Hardy Products is adding new instruments to its research laboratory. Your company has been recommended as a supplier for function generators.

In these two statements, the purpose for the inquiry and the type of inquiry are indicated. In the second part of the letter, the explanation, the writer becomes specific, as shown in the next example.

---

**EXAMPLE**

**Will you please send us the following information concerning the function generator you distribute:**

1. **The cost including shipping costs** _____
2. **The voltage output** _____
3. **The waveshapes** _____
4. **The power requirements** _____
5. **The dimensions and weight** _____

---

Using an itemized list of specific requirements in a letter of inquiry accomplishes three important objectives: First, it clearly designates the type of information needed. Second, it reduces the risk of receiving incomplete data. Third, it permits the reader to save time by filling in the data on the original letter.

A recipient would probably respond to this inquiry even if a paragraph stimulating action were not added. Therefore, the letter may be concluded with a complimentary close at this point, or an action paragraph may be used to suggest that an immediate response is expected, as shown in the following example.

---

**EXAMPLE**

**The underlined spaces following the specific data requested are included to simplify your response. May we hear from you by July 10?**

---

Notice, as in this example, that a specific date is preferable to *soon, now,* or *immediately.*

The three parts of the inquiry letter shown in the foregoing examples are illustrated as a complete letter in Figure 19–1. The full-block format used for the letter in Figure 19–1 differs from the modified-block format illustrated later in Figure 20–1. In the full-block format, all introductory and closing information begins flush with the left-hand margin. In the modified-block format, the date and closing material are indented. A company usually specifies which format it prefers its employees to use.

The letter in Figure 19–1 is an example of format and organization for a letter of inquiry. It also demonstrates how each part of the letter moves logically into the following part. The letter itself, however, does not illustrate a universal writing style. All good letter writers use sample letters as guides but develop individual styles of expression that reflect their own personalities.

HARDY PRODUCTS, INC.
1210 Hunt Road
Hurst, WA  00234

July 1, 1987

Mr. Elliot Frank, Sales Representative
Superior Instrument Company
Street Address
City, State  Zip code

Dear Mr. Frank:

Hardy Products is adding new instruments to its research
laboratory.  Your company has been recommended as a supplier for function generators.

Will you please send us the following information concerning the function generator you distribute:

1.  The cost including shipping costs_____

2.  The voltage output_____

3.  The waveshapes_____

4.  The power requirements_____

5.  The dimensions and weight_____

The underlined spaces following the specific data requested
are included to simplify your response.  May we hear from
you by July 10?

Sincerely,

Ms. Arlene Allen, Supervisor
Research Laboratory

FIGURE 19-1  An example of a positive letter in full-block format

## Writing Acknowledgments

A letter of acknowledgment informs a correspondent that his or her letter, merchandise, or information has been received. Most technicians know when a

Mary R. Silas, Business Agent
Montgomery Corporation
1050 River Street
Denton, Texas  76201

Dear Ms. Silas:

I have just read your letter dated June 1, 1987, requesting an evaluation of the three locations for a new warehouse.

The letter explains that you are specifically interested in three factors pertaining to each location:  (1) the initial cost of land and construction; (2) the annual maintenance costs; and (3) the accessibility to rail, air, and truck transportation.

I'm beginning my research immediately and will have a report ready for you by June 15, 1987.

Sincerely,

**FIGURE 19-2**  An example of a letter of acknowledgment

letter of acknowledgment should be written. Like other positive letters, it usually consists of three parts. The letter of Figure 19-2 shows how these three parts are arranged so that the letter is logical and coherent. The following observations relating to this letter and similar ones are important.

1. Salutations are becoming increasingly varied. *Dear,* for example, is often omitted because, as some people argue, *dear* is not used in conversation. Thus, a simple salutation like the one in the following example is used.

---

**EXAMPLE**
**Mr. (or Editor) Avery:**

---

Writers who initiate correspondence often wonder about an acceptable salutation. Some writers, beginning at once with a conversational tone, open with such terms as *Hello* or *Good Morning.* Others use an attention line: *Attention: Shipping Department.* Still others omit both the salutation and the complimentary close. The letter in Figure 19-2 responds to a letter from Mary R. Silas. Therefore, *Silas* is logically used in the response salutation. *Ms.* is used because marital status is not shown in the original letter and seemed preferable to *Dear Business Agent Silas*, although titles and names are often combined. Sexist terms like *Dear Sir, Dear Madam,* or *Gentlemen* are no longer popular. First-name salutations such as *Dear John* or *Dear Mary* are reserved for business letters between two correspondents who have become well acquainted.

2. The first sentence moves quickly to state that the letter has been read. An unnecessary statement such as "Your letter was received today" is omitted.

3. The second paragraph in the letter restates the details of information given in the letter of request. This restatement is necessary. It verifies that the request is understood and that the reader will comply with the request or will justify any omissions. Second paragraphs in other letters of acknowledgment may ask for additional information or a clarification of some part of the request.

4. If a person writing the letter of acknowledgment wants to emphasize the listed items in the explanation, the writer may arrange them in a column. Double spacing would then separate one item from another.

5. The closing paragraph specifies when the requested information or merchandise will be sent. If necessary, it also specifies how the shipment will be made. This paragraph, then, is not a request for action, as illustrated in a letter of inquiry. It is a promise to act.

## Writing the Good-News Letter

A good-news letter is a letter of approval. The type of work a technician performs determines how frequently he or she can expect to write a good-news letter. For example, a technician who inspects mechanisms that are functioning efficiently writes a good-news letter to the supervisor. A technician may also write a good-news letter to express satisfaction with a company's product. Technicians receive good-news letters, too. For example, a supervisor may approve a new design for a mechanism, a plan for suggested changes, or the purchase of equipment needed by the technician.

Good-news letters are always welcome. People enjoy knowing that something they have said, done, or participated in is successful. For this reason, good-news letters build morale and goodwill. They should be written every time an opportunity for sending them exists.

Like a letter of inquiry or a letter of goodwill, a good-news letter may briefly state, "Your design for a new grinder has been approved. Please submit to the purchasing manager a list of materials needed to construct the experimental model." This kind of note, which can be sent as a memorandum, indicates that the communication from the designer was clear and complete and that no modifications in the design are needed. An explanatory paragraph, therefore, is unnecessary.

In most good-news letters, the writer includes an explanatory paragraph to verify data originally submitted, to ask for additional information, or to suggest some modifications. Then, the message begins, as all good-news messages do,

Dear Pat:

Your suggestion to include a waitress station in the re-
modeled coffee shop is excellent.  Both Harry and I approve the idea and the plans you submitted.

The station will not only improve service but will also be
an attractive addition to the dining room.  Harry wondered,
however, whether an opaque plastic privacy screen would be
more desirable than the carved, pressed-wood screen shown
in the plans.  Do you have specific reasons for prefer-
ring the pressed wood?

Let us know, within the next week, your reaction to Harry's
suggestion so that we can order materials and begin construction.

                                                 Sincerely,

**FIGURE 19-3** An example of a good-news letter

with the statement of approval. The second paragraph discusses any important factors relating to the approval. The closing paragraph may stimulate some kind of action or may end with a pleasant note of appreciation or an offer of assistance.

The letter of Figure 19-3 illustrates a good-news letter in which the concluding paragraph stimulates action. This letter begins with statements of approval. The second paragraph expresses the recipient's favorable reactions before a suggestion to alter the original plan is offered. Criticism or opposing ideas are always more likely to be accepted when they follow good news.

## Writing Goodwill Letters

In business, as in social relations, messages that have only one purpose, to express an interest in other people or companies, are important. A letter expressing sympathy, congratulations, or thanks is always appreciated. It reflects the interest a company has in people as well as in business transactions.

A company or an individual within a company who is considerate of others finds no difficulty determining when goodwill messages are appropriate. Many companies purchase and use cards that express specific messages. Other companies write letters. The letters are usually short. They begin with a statement expressing the main idea. The second paragraph explains the idea in greater detail. For example, a letter of sympathy, such as the one presented in Figure 19-4, expresses sympathy in the opening statement. Because sympathy can be expressed for different reasons, such as sickness, death, or problems created by natural forces, the content of the second paragraph can vary. It may

NATURAL BAKERS
395 Wilson Street
Brighton, MN   53974

November 19, 1986

Kent Barry, Manager
Bestway Bakery
3284 Macon Way
Silverton, OH   49351

Dear Mr. Barry:

We at Natural Bakers have just heard about the disastrous
fire at your plant.  We've tried to call you and will continue to try.  However, we want to be certain you receive
this message expressing our concern.

As you probably know, one of our subsidiaries is located in
your city.  We have asked James Walters, its manager, to
contact you and cooperate in any way that will help you.
It's possible, for example, that you and Mr. Walters can
work out a mutually acceptable agreement for your bakers
to use some of our ovens on a part-time basis.

We hope you are available soon so that we can work on
arrangements that will help you maintain production, at
least on a limited basis.

                              Sincerely,

                              Frances Cole

                              Frances Cole, Manager
                              Natural Bakers

**FIGURE 19-4**  An example of a letter of sympathy

specifically name members of the company who extend sympathy; it may, when death is involved, refer to pleasant associations the company experienced with the deceased; or it may express an offer to help. A letter of sympathy has a tone of hope and cheerfulness; a tone of pessimism is avoided. Perhaps the most valuable part of the letter in Figure 19-4 is its expression of cooperation. Circumstances determine whether a concluding paragraph is needed.

A letter expressing appreciation for information sent by a person is illus-trated in Figure 19-5. This letter includes a third paragraph offering reciprocation.

> Dear Mr. Aimes:
>
> Thank you for the information you sent concerning the surveying course being offered employees in your department.
>
> This information has been valuable to our training department. As training coordinator, I am using the detailed outline you enclosed to develop a similar course for our technicians. With the help you have given, the program promises to be practical, informative, and interesting.
>
> Any time we can help you in a similar capacity or in any other way, please give us an opportunity to do so.

FIGURE 19-5  An example of a letter of appreciation

The greatest danger in writing goodwill letters is that a writer, in an effort to express praise or appreciation, may sound insincere. Insincerity is avoided, however, if the writer objectively evaluates the person involved or honestly appraises the importance of information received. An honest opinion expressed directly never reflects a tone of insincerity.

# SUMMARY

1. The positive letter may be a simple request for assistance, information, or merchandise that a recipient is willing to give. It may transmit good news — such as approval of a request, a plan, or a suggestion — or satisfaction with some service or product. Finally, it may be a goodwill letter that expresses sympathy, congratulations, or appreciation.

2. Because recipients welcome positive letters, they are usually short and easy to write. For these reasons, they are often written carelessly. Guidelines and illustrations in this unit are designed to discourage carelessness and to suggest techniques for preparing well-written positive letters.

3. Most positive letters have three parts: an introduction of the subject, details related to the subject, and a closing statement or two. Each part is illustrated in the letters of Figures 19-1 through 19-5.

4. Of the three types of positive letters, the goodwill letter is the most difficult to write. To simplify the process, a writer must genuinely feel goodwill and develop the skill to express it sincerely.

UNIT **19**

## SUGGESTED ACTIVITIES

**A.** Rewrite the following paragraphs. Eliminate unnecessary words; clichés; incorrect sentence structure; and incorrect spelling, capitalization, and punctuation. You may rewrite the paragraph but do not change the meaning.

  1. As I indicated in our telephone conversation the contract you signed with us on January 21, 1980 will not be considered in effect until such time as all concerned are assured Weekday magazine will be received on schedule and in the quantities contracted for. If after the next issue of Weekday comes on the newsstands your motel is not receiving its copies, please call me collect and we will substitute another magazine for Weekday.

  2. In order to accommodate you and your requirements and in keeping with our company's high regard for unexcelled service, we plan to provide stocks of this new product strategically throughout our present marketing area.

  3. In the very near future, our local sales representative will be in touch with you to explain in detail this new product line which we are pleased to make available to you. In the event you have a need before our salesperson visits your area, we would encourage you to either drop in on us at our local office or give us a jingle and let us give you some ideas we have to help you save some $.

  4. In any event, it is becoming increasingly evident that we must have some vehicle and/or method of providing information to instructors (be they from the regular staff, or outside resource people), about policies and procedures of the extension division. Not only would this ensure equitable treatment of instructors, but it would alleviate a number of problems that arise, due almost entirely to misunderstanding through failure to communicate.

  5. To foster this better understanding and cooperation between our instructors and the division, we propose the compilation of a faculty handbook, in which the policies and procedures of each department in the division utilizing instructors are set forth. We perceive the format of such a handbook to be in the nature of the following:

       1. Introduction and statement of purpose
       2. Home study
       3. Conferences and institutes
       4. Evening courses

**B.** Write a letter of inquiry in which you request specific information concerning a product.

**C.** Write a good-news letter in which a paragraph of explanation is needed.

D. Write an acknowledgment for merchandise, data, or assistance you have received.

E. Write a letter of sympathy to a company that has had a burglary, a flood, or a similar misfortune. Offer specific help that will enable the company to resume normal business more quickly.

# The Persuasive Letter

## OBJECTIVES

After reading this unit, you should be able to do the following tasks:

- Describe the kinds of persuasive letters.
- Write a persuasive request for data.
- Write a persuasive request for equipment.

## OVERVIEW

Persuasion is the act of influencing another person or group of people through reason or logic. In a persuasive letter, then, a reader is influenced to accept ideas or products proposed by the writer. Because the writer anticipates the need to persuade rather than to simply request (Unit 19), he or she must carefully plan and skillfully write persuasive letters.

This unit identifies several uses for persuasive letters. It then illustrates and explains persuasive letters likely to be written by technicians: requests for data and requests for equipment. By combining techniques applicable to all letters (Unit 18) with techniques explained here, a writer can prepare persuasive letters that bring desired results.

# KINDS OF PERSUASIVE LETTERS

Persuasive letters in business and service organizations are written to persuade people to pay a bill; to shop at a particular store; to purchase a particular product; to contribute time, talent, or money to a project; or to comply with a request that primarily benefits the person making the request.

Persuasive letters pertaining to delinquent payments are written by managers of a credit department or by a collection agency. Letters urging people to patronize a store or to purchase a product are written by sales managers. Letters urging a person to give a talk, to act on a committee, to contribute financially to a fund, or to otherwise participate in a project are usually written by men or women in charge of the project.

Most technicians are directly involved in one of two specific kinds of persuasive letters: a letter requesting data or a letter asking for replacement of or addition to equipment. A direct request written when both correspondents profit from the request was included in Unit 19 as part of "Writing an Inquiry or Request." Persuasive-request letters are written when the person making the request is the person who will receive or seem to receive most of the immediate benefits.

# THE PERSUASIVE REQUEST FOR DATA

The art of persuasion is important in all business transactions. People who believe that they can obtain their goals by demands, by threats, or by an "I want, therefore, you will comply" attitude may sometimes gain their immediate objectives. In the meantime, however, they are building reader resentment that ultimately may destroy goodwill and effective communication. The letter of Figure 20-1, written by a graduate college student, illustrates the tone of a request letter that ignores consideration for the reader and concentrates upon I. The single-paragraph format for a three-part letter, the awkward sentence structure, and the incorrect use of capital letters also contribute to the reader's unfavorable reaction to the letter.

A persuasive request for information, like a positive letter, consists of three major parts, as shown in Figure 20-2. The first part is planned to attract the favorable attention of the reader, to interest the reader in the request, and to state the request. The second part justifies the request and shows, if possible, how the reader may benefit by granting the request. The third part specifies the action the reader is expected to take.

## Arousing Reader Interest

When a person asks a favor, that person's first objective is to attract the reader's favorable attention to the request. To accomplish this objective, the writer benefits by employing the "you" attitude and by considering the request from the reader's

```
Dear Language Supervisor:
        I am a graduate student in Classics at Central Univer-
sity and am presently working for a Masters of Arts in
Teaching.  Under the supervision of Professor A.O. Hastings,
I am conducting a survey of Latin courses which are being
taught in secondary schools.  I would appreciate informa-
tion concerning in which years Latin is introduced and
taught, what textbooks are used, and what stress is placed
upon prose composition.  I would also like to know if Greek
is offered at your school.  I would be grateful for any
information or recommendations which you feel are perti-
nent to this survey.
```

**FIGURE 20-1** An example of a poorly written persuasive letter

point of view. As previously stated, writers who are sincere, considerate, and courteous have little difficulty keeping their readers interested; their letters sound confident and convincing.

A technician would not be likely to request reliable data from a person unqualified to give that data. For example, a technician asks about the process for grinding Ferro-Tic from an authority in this process. A technician seeks information about fire control, food service, or cosmetology from experts in those fields. When the opening statement of a letter expresses sincere respect for the reader's special abilities, the reader becomes interested. See the opening sentence in Figure 20-2.

Instead of the statement used in Figure 20-2, a question is sometimes effective, as indicated in the following example.

**EXAMPLE**

**Because your grinding methods for Ferro-Tic have proved so successful, I'm coming to you for help. Will you please answer two questions for me?**

If this question is asked, the letter continues as shown in Figure 20-2. Sending only the attention-getting question accomplishes nothing.

Occasionally, an honest plea for help is effective.

**EXAMPLE**

**Will you please help me? I have two questions concerning methods for grinding Ferro-Tic that you are qualified to answer.**

**NATIONAL REFRACTORIES COMPANY**
**15121 Western Avenue**
**Troy, MS 12345**

February 1, 1987

Mr. Howe R. Miller
Street Address
City, State  Zip code

Dear Mr. Miller:

*Part 1*

You are the one person who apparently has the knowledge and experience needed to answer two important questions concerning methods for grinding Ferro-Tic.

*Part 2*

The information will be used in an instruction manual to be sent to our affiliates in this country and overseas.  We plan, through the distribution of this manual, to help our affiliates appreciate the versatility of Ferro-Tic by learning simpler, more up-to-date methods for machining it.

The two questions I have been unable to find answers for are listed below:

1.  What procedure do you recommend for grinding hardened Ferro-Tic workpieces?

2.  What effect does heat treatment have on the inner diameter of the Ferro-Tic ring?

*Part 3*

The spaces at the end of the questions and the enclosed envelope are intended to simplify your response.  Since our manual must be ready for distribution by March 30, may I hear from you by February 15?

With your permission, your contribution will be acknowledged in the preface to the manual.  Perhaps you would also like a copy of the manual. If so, add a note to this letter, and we'll send you one.

Sincerely,

*Marie Callis*

Marie Callis,
Technical Writer

**FIGURE 20-2** An example of a persuasive request for data in modified-block format

Other opening statements may be more appropriate than the three just discussed. Effective methods for arousing interest result from using good judgment and from understanding the reader. The introductory statement or statements are followed by one or more paragraphs justifying the request.

## Justifying the Request

In justifying the request, as shown in Part 2 of Figure 20–2, the writer is honest and straightforward. She does not digress from her purpose but moves directly into explaining why the request is being made. She explains how the information will be used and, if possible, tells how this use will benefit others, especially the reader. If immediate benefits cannot be indicated, indirect benefits can usually be stated or implied. The reference to the versatility of Ferro-Tic in Figure 20–2 is an implied, indirect benefit to the reader, who is interested in increasing the use of this material.

The writer of the letter in Figure 20–2 is aware that she is, uninvited, seeking the use of her reader's time, energy, experience, and knowledge. Reflecting this awareness while avoiding an apologetic tone is necessary. It is accomplished through skillful writing.

Writers not only use techniques for effective letters when preparing a justification; they also consider the type of business relationship existing between them and the reader, the kind of data being requested, and the circumstances that created the request. The language in the request is not stilted. The tone is relaxed but confident and convincing. Restrictions pertaining to confidential information are respected. Writers do not request information that they know is confidential. Writers pledge secrecy of information that they are expected to keep confidential.

The two questions asked in Figure 20–2 are short enough to be written as part of the letter. If more than two or three short-answer questions are asked, a writer may include a separate questionnaire but should explain the questionnaire in the body of the letter. Additional guidelines for preparing appropriate questions requesting data are listed next.

- Questions must be specific for two reasons: First, a specific question prompts a specific answer, and second, a specific question indicates that the writer knows exactly what kind of information is needed. A vague question often makes the reader wonder whether the effort he or she expends in sending an answer will be appreciated and whether the information will be interpreted correctly.

- Questions are arranged so that adequate space for answers is allowed. The writer must assume that answers may not be typewritten and thus allow space for handwritten answers. The writer is not responsible, however, for anticipating that the reader may send more information than the question asks for.

- The questions must be complete. If dimensions are required, the kinds of dimensions are specified. If the use for the information affects the answer, the use is stated in the question. Required specifications are itemized. A general request for specifications is vague.

- Questions are numbered.

- Questions are worded in a way that permits the reader to write short answers. However, "yes" and "no" answers are seldom satisfactory to either the writer or the respondent.

## Concluding the Persuasive Request

After writers justify their requests, they confidently ask for action. They never assume that the reader will refuse to act. Therefore, they speak in a positive tone and avoid any kind of statement that suggests a cause for refusal. For example, a statement such as "If you, for any reason, cannot send the information, I will understand" and "I apologize for asking you to use your time for my project" is negative and apologetic. It suggests that the writer lacks confidence in his or her ability to communicate effectively.

A request for action, however, is always expressed courteously. A demanding tone must be avoided. The request for action is designed to make action seem easy and to prompt a willing reply. It tells the reader the questions may be answered in the blanks or spaces following the questions. It also refers to the envelope that is always enclosed with a persuasive request for information. After the writer explains what is to be done and how it can easily be done, the writer tells why early action is necessary. A time limit is then specified. See Part 3 of Figure 20-2. The time limit is included so that the reader will not procrastinate. However, the time limit is made to seem less important than the reason for specifying a definite time.

A final statement offering reciprocation, a copy of a finished manuscript that includes the information requested, or appreciation completes the letter.

# THE PERSUASIVE REQUEST FOR EQUIPMENT

A request for equipment written by a technician is usually sent to a supervisor or to a department manager within the company. For example, a laboratory technician requests laboratory instruments or supplies. A machinist may request tools needed for machining new materials. Requests for office furniture, additional work space, new machines, and many other items are frequently written. If the need for the item is obvious to the purchaser and if the cost is fitted easily into the company's budget, a direct request, as explained in Unit 19, may be sufficient. If the item requested is not part of routine purchases, a persuasive request justifying the purchase is needed.

A comparison of the letter report in Figure 13-8 and the persuasive request in Figure 20-3 shows the close relationship in content between the two. However, the

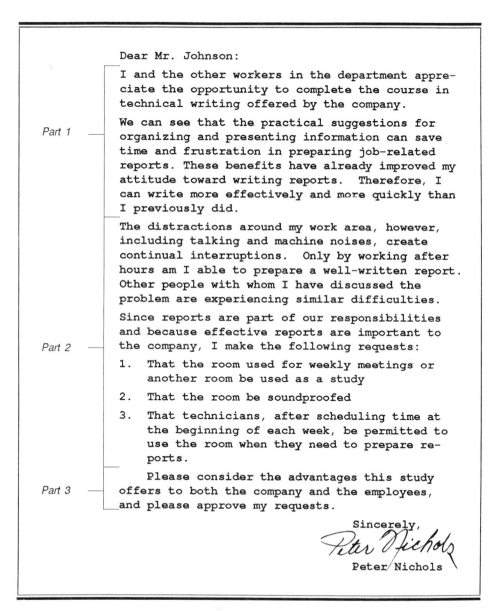

**FIGURE 20-3**  An example of a persuasive request for equipment

letter report is sent in response to a previous oral or written request for detailed information. The persuasive letter initiates the correspondence.

The persuasive letter is usually less formal than the letter report. It is written in a more personal style and expresses the request in a more conversational tone. Also, in contrast to a justification report, the opening statement in the persuasive letter is designed to arouse the reader's attention because a persuasive letter is the initial communication.

Because reasons for requesting equipment vary and because one reader reacts differently from another, a set procedure for arousing interest cannot be established. The technician making the request considers financial benefits to the company, timesaving advantages, and increased employee morale as one or more bases for the opening sentence in the letter. As an example, the opening sentence in Figure 20-3 expresses appreciation. The writer then moves quickly to explain company and employee benefits before making the request.

Since the letter was an intradepartmental communication, a memorandum may have been considered. However, subject lines for persuasive requests are not recommended. Most memorandums include subject lines.

In Part 1 of Figure 20-3 the writer expresses appreciation. He also specifically shows how he and others benefited from help given them by the company. Unlike the request in the letter in Figure 20-2, however, the request follows instead of precedes the justifications. In this position, the request is more convincing

In Part 2, the writer does not specify three requests although he lists three. Omission of the specific number in this letter helps to avoid a demanding tone.

The closing statement of the letter, Part 3, is only a short reference to the request, reasons for the request, and a suggestion for action. Whether an employee should urge action or should specify a time limit for action when writing to a supervisor is doubtful.

A comparison of variations in organization and content of the letters in Figures 20-2 and 20-3 demonstrates again the importance of understanding the reader, knowing the purpose of a communication, and evaluating the circumstances creating the communication before a business letter is written.

## SUMMARY

1. Persuasive letters — requests for information, services, or products that are more likely to benefit the writer than the reader — require skillful preparation.

2. A persuasive letter invades a reader's privacy, primarily for the writer's benefit. Thus, courtesy and consideration for the reader are essential. Also important are the personal attitudes and techniques for writing effective letters discussed in Unit 18.

3. Unless a person writing a persuasive letter can arouse the reader's interest immediately, additional letter content is likely to be useless.

4.  Reasons for the persuasive request must be stated specifically, logically, and clearly enough to convince the reader that the request is justified. The request itself must also be specific.

5.  A person writing a persuasive letter is obligated to make a reader's compliance as easy and simple as possible without sounding submissive.

6.  A persuasive letter concludes with a confident but undemanding request for action by a specified time.

7.  Persuasive letters, like all letters, should reflect a writer's individual style. Stereotyped or insincere expressions can be easily detected and can be nonproductive.

---

## SUGGESTED ACTIVITIES

UNIT **20**

A.  Choose a subject related to your major field of study. Then select a person who is considered an authority in that field but not a person you know personally. Write a persuasive request asking for information. The information may be needed for an oral or written report, for research, or for some other specific purpose.

　　Mail the letter together with an enclosed stamped, self-addressed envelope. Do not specify in the letter that it is a classroom assignment. However, you may state that you are a student conducting research. Make and keep a carbon copy of the letter.

　　Remember that the success of a letter is determined by whether the letter accomplishes the purpose for which it is written. The purpose for sending this letter is to receive a favorable reply. Before sending the letter, therefore, proofread it carefully to determine the following:

1.  Whether the letter begins with an interest-creating statement
2.  Whether the information requested is specifically stated
3.  Whether the justification for requesting the data seems logical
4.  Whether the tone of the letter is (a) demanding or courteous, (b) apologetic or positive, (c) too formally or too informally written
5.  Whether you express consideration for the reader and use the "you" attitude
6.  Whether your letter makes a reply easy
7.  Whether the format of the letter is attractive and correct

Even expertly written requests are not always answered. However, if you do not receive a favorable reply to your request, check your carbon copy to see whether the letter can be improved.

B. Write a persuasive letter requesting a special type of equipment that you, as a technician, would be likely to need. As a classroom project, exchange letters with other students. Evaluate another student's letter according to the rules and suggestions discussed in Units 18 to 20.

C. Many companies willingly cooperate with students in a technical college who are trying to relate theoretical studies with practical applications. If this kind of company is located near your school, ask one of the supervisors for a copy of a persuasive-request letter. The company, after omitting identifying references such as names from the letter, may permit students to use the content of a letter for classroom evaluation. The request for the letter may also be a persuasive-request letter prepared by one or more students.

# UNIT 21

# The Negative Letter

OBJECTIVES

After reading this unit, you should be able to do the following tasks:

- Describe the kinds of negative letters.
- Organize a negative letter.
- Write a negative letter.

## OVERVIEW

A negative letter, sometimes called a letter of refusal, is always written as a response to some kind of request. The person writing a negative letter refuses to send requested information, to purchase requested equipment, to approve credit, to make a claim adjustment, to manufacture or market a new product, or to employ an applicant. A refusal does not always suggest that a request is unjustified. On the contrary, many refusals are written because a person or a company recognizes the request as reasonable but is unable to comply. Nevertheless, a negative letter is always a disappointing message and can thus create resentment, adversely affect goodwill, and result in a series of negative reactions toward the person or company sending the refusal. Any person sending a negative letter avoids or diminishes these reactions by applying the guidelines given in this unit for organizing and writing negative letters.

# ORGANIZING THE NEGATIVE LETTER

Every company wants to build and maintain pleasant relationships with its customers, its competitors, and others with whom it conducts business. Therefore, a company rarely — if ever — sends a letter of refusal without justifiable reasons for doing so. These reasons are important in a negative letter. They give the recipient the opportunity to see that the company is considerate and fair. They also permit the recipient to understand the company's point of view. For this reason, a negative letter is organized into four major parts:

1. The introduction
2. The reasons for the refusal
3. The refusal
4. The conclusions

Each of these four parts is illustrated in Figure 21–1 and explained next.

1. The introduction is sometimes called a buffer because it is designed to reduce the shock of the "no" that must eventually be expressed. Because a negative letter is always a response, the text for the introduction is usually found in the letter being answered. However, the introduction is based upon neutral text or on an idea with which both the writer and reader can agree, as shown in Part 1. The writer, in the first statement, agrees that the company guarantee is still in effect. From this point of agreement, however, he goes directly to the stipulations of the guarantee.
2. In Parts 2 and 3 of the letter, which are combined into one paragraph, the writer explains his reason for refusal. Immediately following the reason, he expresses his refusal quietly but firmly. Although the writer of the letter in Figure 21–1 has only one reason for refusing the request, negative letters contain the number of reasons that pertain to a specific refusal.
3. In Part 4, two paragraphs are written. The first paragraph shows the confidence the engineer feels in his evaluation of the cause for buckling in the concrete. At the same time, he displays a strong "you" attitude by suggesting that the recipient consider other expert opinions. In the final paragraph, the writer again suggests that his analysis is reasonable and fair and that the reader will, therefore, continue to patronize the company.

The type of negative letter shown in Figure 21–1 is one of the most difficult to write. The writer is saying, as courteously as possible, that the reader's request is unjustifiable, but he says it without accusation or without a reflection upon the integrity of the customer. Additional aids for writing this kind of negative letter effectively are given in "Writing the Negative Letter."

A negative letter that is less difficult to prepare and less likely to meet opposition from the reader is illustrated in Figure 21–2. In this letter, the company

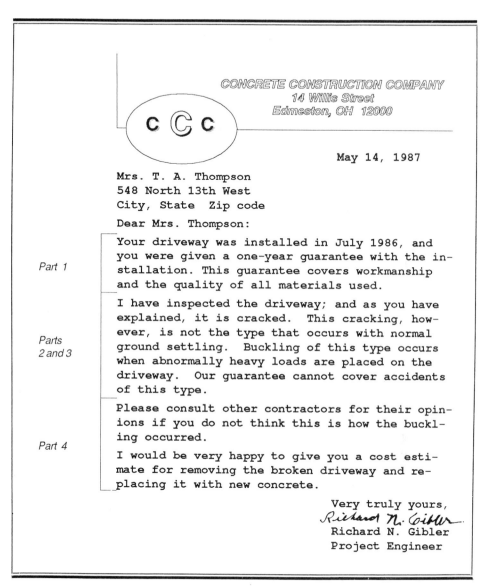

CONCRETE CONSTRUCTION COMPANY
14 Willis Street
Edmeston, OH  12000

C C C

May 14, 1987

Mrs. T. A. Thompson
548 North 13th West
City, State  Zip code

Dear Mrs. Thompson:

*Part 1*

Your driveway was installed in July 1986, and you were given a one-year guarantee with the installation. This guarantee covers workmanship and the quality of all materials used.

*Parts 2 and 3*

I have inspected the driveway; and as you have explained, it is cracked.  This cracking, however, is not the type that occurs with normal ground settling.  Buckling of this type occurs when abnormally heavy loads are placed on the driveway.  Our guarantee cannot cover accidents of this type.

*Part 4*

Please consult other contractors for their opinions if you do not think this is how the buckling occurred.

I would be very happy to give you a cost estimate for removing the broken driveway and replacing it with new concrete.

Very truly yours,

Richard N. Gibler

Richard N. Gibler
Project Engineer

**FIGURE 21-1**  An example of a negative letter concerning repairs

finds no fault with the request but, for reasons beneficial to the person making the request and to the company, decides not to comply. A negative letter similar to the one in Figure 21-2 may be written to an applicant whose qualifications are satisfactory but to whom a job with the company is not available.

In this kind of refusal, the introduction usually states only that the request has been studied and is considered reasonable. This statement, however, must not

May 24, 1986

Mr. William L. Detwiler
3425 Aurora Way
City, State   Zip code

Dear Mr. Detwiler:

The engineers in our office have thoroughly examined your
new instrument for measuring slopes of highways and the
material explaining the design.

Although the instrument is unique and well designed, we
lack the resources necessary to give you the publicity
or the money needed to market your product successfully.
For these reasons, we cannot purchase your invention at
this time.

We suggest that you send your design and materials to the
following companies:

    A.W. Construction Company    K & E Company
    230 South 2nd West          3498 Uintak Boulevard
    City, State   Zip code     City, Street   Zip code

These companies specialize in marketing measuring instruments and can help you much better than we can.  If you
wish, you may use our name as a reference, letting these
companies know that we have great expectations for your
design.  We wish you success in finding a favorable market.

                              Sincerely,

**FIGURE 21-2**  An example of a negative letter of refusal

sound too encouraging. If it does, the reader will be thinking "yes" while the writer is trying to tell the reader "no." Under these circumstances, the refusal may come as a shock.

The letter in Figure 21-2, like the one in Figure 21-1, moves immediately from the introduction to the reasons for refusal, then to the refusal. The fourth part, the conclusion, leads the reader away from the refusal and offers help if possible or expresses continued interest in the reader.

A comparison of the letters in the two illustrations shows that one basic type of organization is used for all negative replies. Specific suggestions for writing all four parts and for adapting information to the type of refusal being written are given in the next section.

# WRITING THE NEGATIVE LETTER

The following sections explain specific techniques that improve the effectiveness of a negative letter.

## Planning the Letter

As stated previously, most people who write a request have convinced themselves that the request is justified. Although the recipient of the request disagrees, the recipient tries to understand the reader's point of view. Consequently, when preparing a negative reply, the writer should observe the following guidelines.

- Use the "you" attitude as the basis for the letter. In this way, the letter shows consideration for the reader.

- Never write in anger.

- Never say anything that belittles or insults the reader.

- Avoid words that seem negative or accusing. Some of these words are *shocked, you claim, surprised, I am unable, you evidently, you stated,* or *amazed.*

- Write confidently, honestly, and sincerely.

Specific suggestions for developing each of the four parts of the negative letter show how the application of these five rules helps a reader accept a refusal without resentment.

## Planning the Introduction

The following three guidelines help a writer prepare an effective introduction for a negative letter.

- The introduction is planned so that no word, phrase, or sentence suggests that the request is being granted. A writer who cannot partially agree with the reader without implying approval can begin with a neutral statement or simply say that the request has been received and evaluated.

- A "no" is never given at or near the beginning of a negative letter. A few people may accept a short reply such as "We are sorry we can't comply with your request." Usually, however, the reader wants to know why. The writer is obligated to explain reasons for refusing.

- The introduction rarely exceeds a single short paragraph. The sentences are arranged so that the last sentence moves the reader logically to the reasons for refusal, as shown in the letter in Figures 21-1 and 21-2.

## Presenting Reasons

The following guidelines may help the writer express the reasons for refusal easily but meaningfully.

- The reasons for refusal immediately follow the introductory statements. The reasons, however, are introduced without an obvious indication to the reader that they are being presented. See the smooth transition to the reason for refusal in the second paragraph in Figure 21-1.

- A writer relies upon the reasons for refusal to maintain rapport with the reader. Thus, the reasons for refusal always precede the "no." A statement similar to "We must refuse your request for the following reasons" is not recommended.

- Reasons are expressed positively, specifically, clearly, and confidently. A positive approach is straightforward but impersonal. Such personal remarks as "You can understand" or "I'm sure you will agree" make the writer sound dictatorial. The reader is not likely to be congenial toward this kind of persuasion. Furthermore, if the reasons are logical, they need no defense. Company policy or a guarantee is not a clear, specific reason for refusing a request unless the purpose for company policy and the terms of the guarantee are explained. Reasons are expressed confidently because the writer, after considering all factors involved in the request, is convinced that the reasons for saying "no" are valid.

- In refusing a new product from an inventor or a designer or in refusing to employ an applicant, the writer is truthful. If the writer recognizes weaknesses in the invention or in the applicant's qualifications, offering constructive criticisms of these weaknesses is usually appreciated by the reader.

## Saying "No"

Saying "no" to a request is always difficult. Four guidelines, however, may reduce some of the disappointment the refusal gives the reader.

- The word *no* is avoided in the statement of refusal. Yet the refusal is expressed in a way that leaves no doubt in the reader's mind that the request has been denied. Frequently, the refusal can be expressed indirectly and still be clear.

For example, "We sell our products only through authorized dealers" refuses the request but avoids a negative statement.

- The statement of refusal should be rather long; long sentences are less forceful than short ones and allow a refusal to develop gradually and undramatically. One effective place to express a refusal is within the independent clause of an inverted complex sentence, as shown in the following example.

**EXAMPLE**

**Because my time schedule does not permit me to leave the city during the week of October 22, I cannot speak before your group.**

If the refusal is written in a separate sentence, it is written as a final statement in the last paragraph explaining reasons. It is never written as a separate paragraph.

- A writer never apologizes for a refusal. The reasons are valid; therefore, the refusal is valid. Thus, an apology is unnecessary and unconvincing. For this reason, negative expressions such as *unfortunately, we regret, we are sorry,* or *we wish we could comply* are avoided.

- An effective writer —knowing that the conclusion is a strong part of a letter— always expresses optimism, interest, cooperation, or appreciation in the final paragraph of any letter he or she writes. Therefore, the refusal is never the concluding statement in a negative letter.

## Concluding the Letter

The contents of a negative letter usually determine the type of conclusion used for the letter. In a refusal to replace damaged merchandise, a sales department often closes with statements related to new merchandise or to a sale on products that may interest the reader. In the types of refusal usually written by technicians, the following guidelines for concluding the letter are helpful.

- A conclusion is designed to maintain goodwill. Therefore, it leads the reader away from the refusal. One of the best conclusions, when it can be used, is an offer to help the reader, as shown in Figure 21-2. This same kind of help is given to an applicant when a company has no jobs available but can suggest the names of other companies to the applicant. A company that does not sell directly to a customer concludes its negative letter with the name and address of the nearest authorized dealer. If help is offered, it must be offered willingly and sincerely. The tone must not suggest that the writer is being benevolent.

- When help cannot be given, the conclusion can express continued interest, appreciation, or both. See the next example.

**We sincerely appreciate the long, pleasant association we have had with your company.**

This conclusion, if used, must evolve logically from the foregoing text. Otherwise, it can prevent rather than encourage continued business relationships.

- The closing paragraph does not give the reader an opportunity to reopen the discussion. Such statements as "If you wish to discuss this further," "If you have other questions," or "If you do not find this explanation satisfactory" encourage continued unfavorable correspondence and indicate that the writer is not convinced that the refusal is justified.

- Closings containing clichés are weak and often sound sarcastic. They are inferior closing statements. Several clichés to be avoided are presented in the following example.

**We hope you find our explanation satisfactory.**
**We know you will agree with our interpretation of the problem.**
**May we hear from you again?**
**I trust that your meeting will be successful.**

- A closing statement such as "If we can be of further help, please let us know" is always avoided. A reader never considers a negative letter helpful. Therefore, the statement seems sarcastic.

# SUMMARY

1. Negative letters are always a refusal in response to some kind of request. Because the person making the request feels it is justified, a refusal is invariably a disappointment. For this reason, a person writing a negative letter needs guidelines that allow him or her to say "no" and still maintain goodwill.

2. Four parts are essential to effective negative letters. Each part is illustrated in Figures 21-1 and 21-2 and explained in the text accompanying each figure.

3. Empathy (Unit 2) is an important quality in a person writing a negative letter. It lets the writer anticipate a reader's reaction to the letter's content and thus diminishes chances of reader resentment.

4. Before writing a negative letter, the writer has concluded that his or her reasons for sending a refusal are valid and, therefore, never apologizes.

5. If a writer includes in his or her letter any help to the reader, the help is offered in a spirit of cooperation. It must not, however, belittle the reader in any way.

6. The writer avoids any statement that invites further controversial correspondence.

UNIT **21**

# SUGGESTED ACTIVITIES

A. Read the negative letter presented in Figure 21-3. Then, list at least five suggestions for writing an effective refusal that have been ignored in this letter.

```
Dear Joe:

This is to acknowledge receipt of your letter.  How well I
remember last year's convention.  We had a great time, es-
pecially on the river trip we took together.  It was a real
thrill when the boat sprang a leak and we had to bail the
water out with a couple of small cans.

Giving a speech before the convention would make me very
happy.  Unfortunately, however, I can't attend the conven-
tion this year.  I'm especially sorry that I can't give
the speech you asked me to give.  You see, I'll be in
Europe that week on a business trip that is very important
to me and my company.  Extend my apologies to the members
of the convention.

Why don't you consider me as a speaker for next year?  In
the meantime, have a great convention, and let me hear from
you soon.  I'd like to hear all about the meetings.  By
the way, if you take another river trip, think of me, won't
you?
```

FIGURE 21-3  An example of a poorly written negative letter

B.  Write a negative letter answering one of the persuasive-request letters in Unit 20. If you prefer, write a negative reply for a persuasive-request letter written by one of the other students in the class.

C.  Arrange to exchange negative letters with another student, and constructively criticize the letter you read according to the suggestions given in this unit and Unit 18. In addition, let your own reaction to the letter influence your critique; the final test of the quality of a letter is always the reader's reaction to it.

# The Resume

## OBJECTIVES

After reading this unit, you should be able to do the following tasks:

- Obtain data for a resume.
- Plan the format for a resume.
- Organize the data for a resume.
- Write a resume.

## OVERVIEW

A resume, also called a data sheet, is the most important aid an applicant can have. It is used for personal interviews and for filing applications with most employment agencies. It is also used as an attachment to letters of application sent to possible employers. A foresighted technician keeps a copy of a well-written resume in a personal file and alters it periodically as his or her education and work experiences change.

When a resume and a letter of application are sent to a company, the resume is considered a supplement to the letter. The resume, however, is always prepared before the letter of application is written. Therefore, the resume is explained in this unit. The letter of application is explained in Unit 23.

A resume is an outline of an applicant's history. It may include five different types of data: personal data, education, work experiences, personal interests, and references. In recent years, career goals of the applicant are also included.

There are no strict rules governing what to include on or how to arrange a resume. Therefore, this unit is limited to suggestions that will help technicians prepare effective resumes. The topics discussed are obtaining the data, planning the format, and organizing the data.

# OBTAINING THE DATA

The primary purpose of a resume is to interest the employer in the applicant's qualifications for a particular kind of work. Therefore, the best resumes are written by people who have carefully analyzed their qualifications in relation to the kind of work they want to do. To make a useful analysis, the applicant places each of the five types of data mentioned previously — personal data, education, work experiences, personal interests, and references — at the top of separate sheets of paper. Each type may later become a major heading on the resume. In addition, the applicant's career goals are clarified in a written statement. Even if this statement is not used later in the resume, it is an excellent guide for the content of the letter of application. The following comments explain the data that should be included on these five or six sheets.

1. On the personal work sheet, the applicant's age, height, weight, health, and marital status are listed. Some people feel that such information is not revelant to one's ability to perform on the job, and they do not include a personal-data section on their resumes.

2. On the education page, the names and addresses are listed for all the applicant's schools beyond high school. The person's major subject and average grade, if 3.0 or above, are also listed. In two separate columns, the applicant then lists all important courses of study: subjects directly related to his or her specialization in one column and subjects indirectly related but contributive to his or her specialization in the other. A complete list of all academic courses by years, semesters, or quarters is not given. The employer probably will not be willing to take the time to read such a list to find the courses that are applicable.

3. The applicant next lists all work experiences, beginning with the writer's current or latest employment. Under each job, the length of time in months or years that the person was employed is noted, and the kind of work involved in each job is specifically stated. Students completing two-year technical college may have only part-time work experiences. These jobs are included as shown in Figure 22-1.

4. Some applicants see little value in including personal interests and activities in a resume or in discussing them in a letter of application. Surveys show, however, that knowing a person as well as the person's skills helps a personnel manager find more suitable employment for that person. Therefore, on the personal-interests sheet, the applicant should subdivide the major heading into activities, interests, and, if applicable, honors. Under activities, the club, church, charitable, or community groups in which the applicant actively participates are listed. The type of participation involved is also specified. For example, one may sing in a chorus, teach Morse code to Boy Scouts, or be an officer in a club. Under interests, hobbies developed for personal enjoyment — such as fishing, camping, photography, or ham radio — are listed. A person

<u>Qualifications of Marta N. Allen</u>
<u>for Secretarial Position at Ideal Chemical Company</u>

<u>Home Address</u>
211 Northrup Drive
City, State  Zip code
Telephone  111-2345

<u>Career Goals</u>

To develop the efficiency, personality, and understanding
of the operation of a firm that will enable me to assume
responsibility and be held accountable for it; to solve
problems, and to advance professionally and economically.

<u>Education</u>

Graduated from Hales Secretarial College, Des Moines, Iowa,
June 1986

| | |
|---|---|
| Business-related Courses: | Shorthand (120 words per minute), typing (80 words per minute), business machines, office procedures, business communications, principles of management, filing, secretarial problems |
| Personal-development Courses: | Sensitivity training (group sessions), business behavior, personality development, industrial psychology |
| Grade-point Average: | 3.67 on a 4-point scale |

<u>Work Experience</u>

Two years in part-time general office work at Valley Medical Center.  Worked at switchboard and reception and appointment desks.  Also scheduled surgery and did occasional typing.

Earned extra money during high school as part-time waitress
at Mary's Restaurant and as baby sitter.

Born  3/12/65                Hobbies include piano playing,
Single                            tennis, gardening
Excellent health           Active in Girl Scouts, church,
                                        and tennis club

<u>References</u>

Claude L. Forest, Professor       Mrs. Mavis Aron, Counselor
Hales Secretarial College         Hales Secretarial College
299 West First Street             299 West First Street
Des Moines, Iowa 52210            Des Moines, Iowa 52210
Telephone: (515)111-7234          Telephone: (515)111-4567

Mrs. Sara Martin, Supervisor      Mr. Arthur Steele, Director
Valley Medical Center             1770 Hillsdale Drive
1616 Marshall Street              Davenport, Iowa 52805
Davenport, Iowa 52805            Telephone: (319)111-5678
Telephone: (319)111-3456

FIGURE 22-1  An example of a resume

who has received awards or honors lists them as a third part of personal interests.

5. On the references sheet, the applicant lists names, addresses and telephone numbers of about six people who can give an honest appraisal of the applicant's work, educational achievements, and personality. Before these names are used on the resume, however, the applicant receives permission from the person being named as a reference.

6. Finally, a statement of career goals specifying the applicant's immediate and long-range plans is written. This statement, often helpful to both the applicant and the employer, should indicate at least a desire to improve personally and a desire to contribute toward future company development.

After the five (or six) pages of information are completed, the applicant is prepared to write the resume for a particular organization. As mentioned later in Unit 23, an applicant improves the chance for success if he or she learns as much as possible about the potential employer. This knowledge helps determine how detailed the resume should be. For example, some organizations want little more than name, address, education, and work experience. Once the applicant has selected appropriate data, he or she organizes it, preferably on a single page. The original work sheets are filed for future reference.

# PLANNING THE FORMAT

Resumes vary in format; but in all cases, the information must be attractively and logically presented. Logical major headings and subheadings are the most important aids the applicant can use in preparing a resume that will appeal to the reader. One widely accepted format is illustrated in Figure 22-1. Others are shown in various business-related textbooks, including those listed in the supplementary readings following Unit 23. The following guidelines are offered to help applicants prepare attractive and informative resumes.

- All resumes are typewritten or printed on the same kind of paper used for business letters and reports (see Unit 7). If text is typewritten, a new letter (not a carbon copy) is prepared for each potential employer. A carbon copy suggests that an applicant is indifferent about obtaining employment.

- The title of the resume may be written in either of two ways: The first letters of all important words are capitalized and the title is underlined; or the entire title is capitalized.

- Photographs, once extensively used on resumes, have lost their popularity. If, for any reason, a photograph is used, it may be located three spaces below the title of the resume, placed on the left-hand side of the page, on the right-hand side of the page, or in the horizontal center. The centered photograph is the most widely used. The applicant is photographed in a straightforward pose. His or her clothing is conservative, business apparel. The photograph is unretouched. Its size should be approximately 2 inches by 2 1/2 inches.

- The applicant's address and telephone number are listed. Figure 22-1 shows one format for listing them. Others are shown in several business-related textbooks.

- The typewritten text is surrounded by white space. If two lines are needed for the title of the resume, the title is double-spaced (Figure 22-1). Three spaces separate the title from the rest of the data. The title begins no less than 1 inch from the top of the page. A 1-inch margin is also allowed on the sides and on the bottom of the page. (The left-hand margin is reduced to 1 inch because a resume is seldom placed into a folder or bound.)

- The page is arranged so that it appears balanced. The five or six major headings explained in "Obtaining the Data" may be either centered or placed on the left-hand margin. Information listed under a left-hand major heading is indented five spaces, as shown in the following example.

**EXAMPLE**

**Education**

     **Graduated from Helmstead College, Albany, New York, June 15, 1979**

The major headings, like the title, may be written in one of two ways: Only the first letter of all important words is capitalized and the heading is underlined, as shown in Figure 22-1; or the entire heading is capitalized. Headings, however, are entirely capitalized only when the title is entirely capitalized.

- Subheadings used under centered major headings are flush with the left-hand margin. Subheadings used under major headings written on the left-hand margin are indented five spaces. Two spaces separate subheadings from major headings.

- Text relating to a major heading or subheading is single-spaced. A double space separates one group of information from another. See information under "Work Experience" in Figure 22-1.

- If used, references (four to six are recommended) are usually listed in two columns horizontally balanced on the page. The information concerning each reference is single-spaced. A double space separates one reference from another, as shown in Figure 22-1.

- A resume sometimes extends beyond one page. One inch from the top of each succeeding page an abbreviated title and the page number are written, preferably on a single line.

---

**EXAMPLE**     **Marta N. Allen's Qualifications for Secretarial Work, page 2**

---

# ORGANIZING THE DATA

Two general kinds of resumes are possible. One is for the applicant who is not seeking a specific kind of work. The title of this resume may be "Resume," "Personal Data Sheet," or a similar general title. A more effective title, as shown in Figure 22-1, specifies the kind of work the applicant is seeking. This title, like titles for formal reports, is usually long but is informative.

Personal data has conventionally been the first item on the resume. Qualifications for employment, however, are not determined by a person's physical characteristics. For this reason, an increasing number of applicants list personal data near the end. Some include it with activities and interests under the major heading "Personality."

The statement of career goals is a relatively new addition to resumes. If an applicant decides to express such goals, they are usually the first major entry. "Career Goals" or "Objectives" is the major heading, as shown in Figure 22-1.

The two most important parts of a resume, education and experience, are arranged in their order of importance. A college or high school student probably relies more upon education than upon experience to obtain employment. If so, information related to the person's education precedes that related to work experience. (As a person gains experience, these two items will probably be reversed on his or her revised resume.)

"Education" is a major heading. Under this heading, the following information is given:

—The names and addresses of schools the applicant attended
—The degree or degrees the applicant received
—The month and year the applicant graduated
—The applicant's average grade such as A, B+, or B; or 3.85, 3.25, or 3.0

An average below B or 3.0 is omitted. An applicant who has attended more than one college lists first the one from which he or she received the highest degree.

Selected classroom courses, discussed in "Obtaining the Data," are listed under appropriate subheadings. For example, a drafter may list all completed drafting courses and those closely related to drafting under such subheadings as "Work-Related Courses" or "Courses Related to Drafting." Other courses that help the employer understand the applicant or that have an indirect relationship to the work being applied for are grouped under a second subheading such as "Other Technical Courses," "Personal-development Courses," or some other appropriate subheading. See the two subheadings under "Education" in Figure 22–1. A brief explanation of the content of a course, if significant, is given after the name of the course. The explanation, like all information on the resume, is condensed. Specific words and phrases replace sentences, as indicated in the following example.

**EXAMPLE**

**Shorthand (120 words per minute).**

**Report writing (emphasis upon writing techniques).**

Work experiences are listed under an appropriate major heading such as "Experience," "Work Experiences," or "Work Experiences in Geology." Jobs are listed chronologically from the latest job to the earliest one. However, the list seldom includes more than the last five jobs unless an earlier job is directly related to the work currently being sought.

The work list specifies the years during which each job was performed, the title of the job, the place of employment, and the specific kinds of activity involved in the job. The applicant concentrates upon activities in a previous job that qualify him or her for the job. In the following example, the applicant wanted to become a sales trainee for a major oil company. The explanations of his or her previous jobs show that the person is familiar with the oil industry.

**EXAMPLE**

**1980–**        **Seneca Oil Company, Akron, Ohio.**
              **Worked in the laboratory as a No. 2 tester.**
              **Also worked in the Gas Recovery Plant.**

**1978–1980**   **Joe's Service Station, Akron, Ohio.**
              **Part-time service station attendant.**

Anything that helps an employer understand an employee's personality is listed under the major heading "Personal Interests." "Activities," "Interests," and "Honors" are three possible subheadings. Statements relating to each activity, interest, or honor must be brief. However, factors that indicate an interest in people, a competitive nature, creative ability, an interest in detail, or other personality traits should be reflected in these statements.

Personal data may simply be given in a list at or near the bottom of the resume (see Figure 22–1). Or a heading such as "Personal Data" (conforming to the format for other headings in the resume) may be used. This information, however, precedes a list of references if one is included.

"References" is always the last entry on a resume. As stated previously, an applicant receives permission from a person before listing that person's name as a reference. Seeking the person's permission is an act of courtesy and also guarantees that the person feels qualified and is willing to give a recommendation.

# SUMMARY

1. Resumes, occasionally called data sheets, are primarily an applicant's work and education history.

2. Resumes are submitted with an application letter or a cover letter when mailed. A person who has not submitted a resume before being interviewed often hands it to the interviewer.

3. Resumes vary extensively in format and content. Many acceptable ones can be found in the business-related texts listed in the Suggested Supplementary Reading list, page 275. An applicant prepares the resume most likely to appeal to a potential employer.

4. Any format chosen should be attractive and easy to read. It should also be organized so that the reader focuses upon the applicant's name, experience, and education, usually the primary criteria for employment.

5. An applicant's address and telephone number are always given with or near the applicant's name at the top of the resume.

6. Other data, including personal data—such as birth date, marital status, and health; personal interests, honors, and activities; and names of references — may or may not be included. When they are, they are placed in a secondary position — at or near the end of the resume.

7. Preferably, a resume uses only one sheet of high-quality paper. It is typewritten or printed. Appropriate headings identify the primary data and references. A heading may or may not precede personal data.

8. Students who prepare and periodically update a complete resume, including secondary data, simplify the possible ongoing process of submitting one.

UNIT

# 22

# SUGGESTED ACTIVITIES

A. In the school or public library, find other textbooks containing examples of resumes. Some are listed in the supplementary readings. In a classroom discussion or as an individual project, compare these resumes with the one shown in Figure 22-1. Then, select a format that can be effectively used for your resume. Make an outlined sketch of the format to be used later.

B. Select appropriate major headings for your individual resume work sheet; then, organize all information that will help you obtain employment under these headings. When necessary, group the information under two or more subheadings. Subheadings in a resume duplicate subtopics in a report. Using this information as a guide, determine the best location in your resume for each type of information. Then, number your major headings according to this arrangement. For example, if "Education" is to be entered first on the resume, identify "Education" as No. 1. Also, select an appropriate title for your resume.

C. A person who plans to seek employment may need many original typewritten resumes. If so, a backing sheet is useful. The backing sheet shown in Figure 7–1, used for formal reports, is not suitable for resumes. Prepare a new one. (1) Mark a 1 1/4-inch margin from the top of a standard-size sheet of typing paper, (2) mark a 1-inch margin from the bottom and from the right- and left-hand sides of the paper, (3) outline this margin heavily in black ink, and (4) draw a black vertical line through the center of the paper. This centerline simplifies typing the title of the resume and all center headings. The margin lines indicate the margins for other data.

D. Using the backing sheet, prepare a resume that is specific, meaningful, easily interpreted, and attractive. Do not consider this activity as merely a classroom exercise. Plan a resume that can be used any time you send a letter of application, file an application with an employment agency, or accept an opportunity for a personal interview.

UNIT

# The Letter of Application

## OBJECTIVES

After reading this unit, you should be able to do the following tasks:

- Plan and organize a prospecting letter of application.
- Plan and organize an invited letter of application.
- Write a letter of application.

## OVERVIEW

The resume discussed in Unit 22 and the letter of application (or cover letter) complement each other. Therefore, when an applicant seeks employment through written communication, the resume is always attached to the letter of application. The resume lists the products—education, experience, and personality —that an applicant is offering. A letter of application explains how a company or organization will benefit by using these products. Thus, an employer's favorable or unfavorable response to a letter of application often depends upon how effectively the applicant coordinates his or her products with the company's needs. This unit illustrates and explains two different kinds of applications: a prospecting letter and an invited letter. It also offers suggestions to help applicants write letters that appeal to a potential employer.

# PLANNING AND ORGANIZING THE PROSPECTING LETTER

Prospecting letters are letters written by applicants who know the kind of work they want to do and the kind of company for which they want to work. However, they do not know whether positions are available. Such applicants write letters explaining the jobs they want and their qualifications. They also specify that they are applying for work. By means of carefully planned letters, they hope to receive favorable replies.

## Planning the Content of the Prospecting Letter

In a prospecting letter, an applicant tries to include six kinds of information, as illustrated in Figure 23-1. The marginal numbers in Figure 23-1 correspond to the six types of information discussed next.

1. The applicant indicates that he is familiar with the work performed by the company to which he is applying by relating his qualifications to the company's requirements. Methods for showing this relationship are explained later in this unit. An example is shown in No. 1 of Figure 23-1.

    To learn something about the company, an applicant can talk to counselors, to employment agencies, and to employees of the company. An applicant can write to the company for any available literature. An applicant can also check technical libraries for information or take advantage of tours routinely conducted by many companies.

2. The applicant in the example not only expresses an understanding of a company's operations. He also expresses a sincere interest in the company and states that he wants to work for it. See No. 2 in Figure 23-1. Expressing interest is not difficult for the person who knows why he or she decided to apply to this particular company. Such qualities as a company's stability, its prestige, its flexibility, its integrity, or its product excellence may be appreciated by an applicant.

    One successful applicant expressed interest in a petroleum company by saying, "I would be delighted to join your organization especially because it is a leader in petroleum products." Another applicant showed knowledge of and interest in a company in the following statement: "Because we share an interest in pollution-control equipment and systems, I want to work for your company."

3. An applicant usually specifies a willingness to relocate and to travel. See No. 3 in Figure 23-1. This requirement in a prospecting letter may seem demanding, and perhaps it is. Yet a realistic study of industry shows that many companies have subsidiaries and affiliates in nearly every part of the world. Communication with these organizations is essential to company

244 Sunset Drive
Rockford, Illinois 61107
December 14, 1987

Ms. Michelle J. Bonds, Director
Personnel Department
Intercity Gas Company
1495 South 250 West
Denver, Colorado 80216

Dear Ms. Bonds:

*No. 2*

The increasing demand for natural gas as a fuel in the home, in industry, and in electricity production indicates that your company has stability and growth potential.

I want to use my technological training to contribute to that growth. Therefore, I am submitting this application for the position of drafter.

*No. 4*

As indicated in the attached resumé, I have completed the requirements needed for a drafter.

*No. 1*

In addition, I have completed one year of college chemistry. This knowledge of chemistry helps me understand the importance of natural-gas by-products such as carbon black, plastics, solvents, and gasoline.

*No. 4*

Other courses such as mechanical technology, physics, and calculus shown on the resumé indicate my continued interest in all subjects related to technological development. Because of this interest, I will continue to attend school at night while I am employed. Both your company and I should benefit from this additional education.

*No. 5*

In aptitude tests given at the University of Rockford, I was rated in the ninetieth percentile in abstract reasoning. This test indicates my interest and potential in graphics and engineering. Interest, I believe, breeds hard work and opens the door to advancement.

*No. 3*

Of course, if your company requires it, I am willing to relocate or to travel.

*No. 6*

I can be at your office any time that is convenient for you to discuss my qualifications personally. My telephone number is 815-482-9710. I am usually available until 11 A.M. daily.

Sincerely,

*James Heroldson*

James Heroldson

**FIGURE 23-1** An example of a letter from an applicant prospecting for work

growth. In addition, conventions, shows of new equipment, and courses in new techniques are held in various parts of America and in foreign countries. A company competing in the rapidly changing field of technology must know what advances others are making. It must also efficiently use the skills of people it has helped to train. Therefore, most technicians will be expected to travel and possibly to relocate.

Some exceptions to including a statement regarding relocation and travel must be considered by an individual applicant. For example, a beginning stenographer, welder, or auto mechanic is not likely to travel. A person applying for work in a small local organization would not expect to be relocated.

4. A resume is attached to every letter of application. The relationship between the resume and the letter of application is explained in the introduction to this unit. An example is shown in No. 4 of Figure 23-1.

5. If possible, some special personal trait that distinguishes the applicant from other applicants is shown. In No. 5 of Figure 23-1, the applicant relates abstract reasoning to successful performance. A similar example is shown in Figure 23-2.

6. The applicant requests a personal interview. Some employers want applicants to say they will come for an interview at any time that is convenient for the employer, as illustrated in No. 6 of Figure 23-1. Many letters of application, however, designate dates and times when the applicant is available, as shown in Figure 23-2. Therefore, a conflict exists between two points of view. Applicants, after considering both alternatives, must decide whether they can or want to be always available for interviews or whether they prefer to specify certain times.

## Organizing the Prospecting Letter

Letters of application are persuasive letters, almost identical in organization to the persuasive letters explained in Unit 20. Applicants are persuading companies to pay them for the use of their skills. They, therefore, must arouse the reader's interest, persuade the reader that the company the reader represents needs the qualifications listed in the resume, and encourage the reader toward favorable action.

In the first two paragraphs in Figure 23-1, the applicant arouses interest by talking favorably about the company. He also tells the specific kind of work for which he is applying. Interest may also be aroused by mentioning the name of a person known to the reader. For example, an opening statement may say, "Mr. Harold Brown, your field representative, suggested that I apply to you for work as an electrician." Applicants may begin prospecting letters by mentioning qualifications they possess that they know or think are needed by the company. For example, one may begin by saying, "Because of my training and experience in electromechanisms, I believe my services can be valuable to your company."

211 Northrup Drive
Davenport, Iowa 52807
July 19, 1987

Mr. Robert Eams
Personnel Manager
Ideal Chemical Company
1975 Southward Road
Davenport, Iowa 52803

Dear Mr. Eams:

I believe my training, experience, and personality qualify
me for the position of receptionist and stenographer listed
in the Evening Herald, July 18, 1987. I, therefore, am
submitting my application.

As indicated on the attached resumé, I graduated from Hales
Secretarial College in June. Before graduation, I developed a shorthand speed of 120 words a minute and a typing
speed of 80 words a minute.

From my two years' experience at Valley Medical Center and
from classes in personality development, I have learned to
communicate effectively both in person and on the telephone. At the Medical Center, I also learned to type information containing difficult medical terms.

The personal-development courses taken in college and listed on the resume are valuable for one other important
reason: I have learned to accept constructive criticism
without becoming personally offended. This quality, I believe, will help me continually improve as a stenographer
and as a receptionist.

Please call me at 824-1116 and give me an opportunity to
talk to you personally. I am home every day until 2 P.M.

Sincerely,

Marta N. Allen

Marta N. Allen

FIGURE 23-2  An example of a letter of application for an available position

The persuasive part of the letter usually consists of two or more paragraphs.
The length is partly determined by the years of training the applicant has had, the
number of years the person has worked, and the types of work the applicant has
done. However, an applicant who wishes not to bore the reader learns to be

selective in correlating only significant parts of a resume with a particular job. The technician who wrote the letter in Figure 23–1 used three paragraphs to show how his experience, training, and personality would benefit the company. All his information was specific, interesting, and relevant.

No specific guidelines can be made for organizing persuasive information. However, important material is preferably placed at the beginning of a paragraph, and each paragraph should develop only one major idea. The entire letter should be unified, coherent, and logically developed.

The third part of the letter, requesting favorable action from the reader, always expresses the applicant's desire for a personal interview. Arranging the interview should be made easy for the reader. If the applicant can be reached by telephone, a telephone number is given. If the area code is needed, it is included. If the applicant can be reached by mail, a complete address including the zip code is given. Preferably, both the telephone number and address are included. The applicant also specifies whether he or she can be available for an interview at any time or whether, because of other commitments, the applicant is available only on certain days or at certain times. However, an applicant who makes arranging an interview seem difficult is not likely to be called.

Most employment counselors do not recommend that applicants enclose a return envelope with their applications. They believe that an employer is willing to pay for telephone messages or mail sent to a potential employee.

# PLANNING AND ORGANIZING THE INVITED LETTER

Often, a company advertises for employees through newspapers, technical journals, or other publications. An applicant responding to an advertisement is writing an invited letter of application. A company advertising a job is looking for a person whose qualifications meet the company's needs. Through the advertisement, the company reaches a large number of potential employees.

## Planning the Content of the Invited Letter

Before writing invited letters, applicants study the advertisement to determine whether they are fully qualified for the job being offered. If they cannot meet all requirements, they probably should not apply. An invited letter that is submitted contains the following information:

1. One or more statements telling where and when the advertisement was published and the kind of work it specified.
2. A statement indicating that the letter is an application for the job.

3. A correlation of the information in the attached resume with the kind of work and any other qualification specified in the advertisement.
4. A request for an interview.

Unless applicants are confident that additional information will help them obtain the job, they associate everything written in the invited letter with the text of the advertisement. An example of an invited letter is shown in Figure 23-2. The information for the letter was taken from the sample resume of Figure 22-1.

## Organizing the Invited Letter

The invited letter is usually shorter than a prospecting letter because the invited letter does not include statements expressing interest in or knowledge about a company. The two different kinds of letters, however, have identical organizations, as shown in Figure 23-2. In contrast to the prospecting letter, the invited letter begins with a statement referring to the advertisement being answered. The applicant first tells where and when the advertisement was published and the kind of employment it offered and then states that he or she is submitting an application.

The organization of the second and third parts of an invited letter is the same as that for the prospecting letter. Only the content differs. Applicants rarely say anything about the company or the company's policies. They do not mention their willingness to relocate or travel unless the advertisement indicates the need. However, applicants do emphasize any job-related personal trait, training, or experience that distinguishes them from other applicants.

# WRITING THE LETTER
# OF APPLICATION

The following guidelines help applicants establish and maintain effective communication with their readers.

- Applicants use positive statements. They are confident that they are qualified to do the work and willing to work. They express their information directly, specifically, and convincingly. They do not suggest, by negative ideas, that their applications will be refused.

- Applicants' personalities are reflected in the way they write. Therefore, they compose their own letters of application. They never alter a sample letter to adapt it to their qualifications.

- Applicants avoid stereotyped or stilted expressions and sentences. They write to prospective employers in a natural, easy manner.

- Applicants never sound demanding. As in all persuasive requests, they are asking for a favor. But at the same time, they should not be apologetic or submissive. They expect to repay the favor by using their abilities to benefit the company.

- One survey of employers indicated that they expect two things when they hire a person; first, that the person be intelligent enough to learn the job and, second, that the person be flexible enough to adapt to changes that occur in all dynamic companies. Any tone or wording in a letter of application that reflects these traits in an applicant is effective.

- The "you" attitude should dominate the letter. However, obvious omissions of "I" make the letter sound weak and insincere.

- Applicants do not overestimate their training and experience. They avoid a tone that suggests "Let me tell you how good I am." One student said to a professor in a college communications class, "I really can't see why I should learn any more English. I figure I already know as much as any of my future bosses will." Such an arrogant attitude will probably be reflected in any letter this person writes.

- Applicants are sincere and honest. They never try to conceal physical weaknesses or weaknesses in training. By admitting the weaknesses and showing how they compensate for them, applicants indicate sincerity and honesty that an employer usually appreciates. Sincerity is also important when applicants compliment a company. Compliments that are not genuine are omitted.

- Applicants try to avoid references to salary in letters of application. If an advertisement specifies that applicants name a salary, they can usually avoid exact figures. An expression such as "your usual wage scale" usually satisfies the employer.

- Applicants do not waste space and lessen their readers' interest by using unnecessary expressions such as "I read your ad," "I would like to apply (or be considered as applicant)," or "As you can see from my resume." They also avoid general terms such as *invaluable, outstanding, superior,* or *more than qualified.*

- Applicants never send duplicate copies of letters of application to employers. They type each letter separately. In fact, they probably alter some parts of each letter so that the content is adapted to the needs of a specific company.

- Applicants use the high-quality typing paper described in Unit 7. They never use company stationery. They arrange their format as explained in Unit 8.

Preferably, they place their letters and attached resumes in envelopes that measure 9 1/2 inches by 4 1/8 inches. The envelopes should match the paper used for the letters. They sign the letters legibly in black or dark-blue ink.

# SUMMARY

1. A letter of application, also called a cover letter, complements a mailed resume (Unit 22). This letter is important. It extends to a potential employer the personal attention not shown in a resume. It also provides an opportunity for the applicant to explain how his or her individual talents and attitudes fit the needs of a particular organization.

2. A letter of application is brief, preferably limited to one page.

3. The letter does not duplicate data given on the resume; it refers the reader to the resume. However, it can explain how one or more important qualifications in the resume relate to the job being sought.

4. Letters of application are persuasive letters. They are organized, therefore, to arouse the reader's interest, to persuade the reader to offer employment, and to encourage action (See Unit 20).

5. Letters of application are divided into two general classes: the prospecting letter and the invited letter. Each is defined, explained, and illustrated in this unit.

6. Applicants who research companies and indicate in their letters a knowledge of and interest in a particular company write the most effective letters of application. Therefore, the content of a letter sent to one company rarely duplicates that sent to another.

7. Letters of application should project sincerity, honesty, self-confidence, and individuality.

UNIT **23**

# SUGGESTED ACTIVITIES

A. All readers do not react in the same way to a particular letter. The invited letter of application presented in Figure 23–3 was rated as "good" by one reader. Yet the letter violates nearly all suggestions given in this unit for writing effectively. In a classroom discussion or as an individual, analyze the letter. By using specific references from the letter, explain your reaction to the following:

   1. The tone of the letter

Dear Mr. George:

My experience and record in managing a large nonferrous metallurgical plant may be of interest to you in your search for a Superintendent of operations that was noted in ad in the Engineers' Journal on June 15.

I have for eight years been Plant Manager of a large smelter employing 850 people.  As top administrative official for this facility, I have been responsible for its operations, technology, labor relations, and cost control policies.

During the period I have held this position, labor productivity has increased over 100 percent.  Total production cost per unit of product has decreased 19 percent in spite of a 28 percent increase in wage rate and in spite of other escalations.

I hold several patents that are in commercial use in the industry and am the author of several published technical papers relating to metallurgy.

After graduating with a B.S. degree in chemical engineering, I joined the navy for four years.  I then took a year of graduate study in metallurgy at the Northeastern University and in 1962 entered the employ of Smith & Allen Corporation as Testing Engineer, becoming Plant Manager in 1980.

I am 40 years old, in excellent health, married, with three children ages 14, 12, and 10.  I am active in church and civic organizations.  I am also a member of the National Society for Engineers.

I am planning to be in San Francisco on or about July 11 and could review the matter of possible employment with you if my qualifications interest you.  I could at that time furnish names of persons well known to you who can testify to my professional competence and good character.

Sincerely,

*John F. Allisen*

John F. Allisen

*Achievements*

FIGURE 23-3  An example of an invited letter of application

2. The opening paragraph
3. The paragraphs explaining the applicant's qualifications
4. The closing paragraph

B. Ask one or more personnel managers from local companies or service organizations to discuss letters of application with members of your class. Ask the speakers specific questions relating to the subject matter in this unit. Keep notes of the discussions. Then, decide the kind of information you want to include in your letters of application.

C. Using the information in the resume prepared for Unit 22, write a prospecting letter of application.

D. In a local newspaper or other publication, find an advertisement that offers a job for which you are qualified to apply. Write an invited letter of application. Keep carbon copies of all letters you write. They are always useful as guides for letters you may want to write later.

SECTION V

# SUGGESTED SUPPLEMENTARY READING

Damerst, William A. *Clear Technical Reports*. New York: Harcourt Brace Jovanovich, Inc., 1972. On pages 170 to 174, the author discusses many words that affect the tone of a letter. Various kinds of letters are identified and illustrated on pages 157 to 178.

DeVries, Mary A. *The Prentice-Hall Complete Secretarial Letter Book*. Englewood Cliffs, N.J.: Prentice-Hall, Inc., 1978. Chapter 8 explains and illustrates several appreciation and goodwill letters. Forms for identifying correspondents in inside addresses and salutations are given on page 257. These forms, when used for various government, court, foreign, church, and university officials, are arranged in tables on pages 258 to 280.

Houp, Kenneth A., and Thomas W. Pearsall. *Reporting Technical Information*. New York: Macmillan Publishing Company, Inc., 1984. Letter formats are described on pages 279 to 288; letters of inquiry, page 291; letters of application, pages 303 to 306; and resumes, pages 305 and 307 to 311.

Lannon, John M. *Technical Writing*. Boston: Little, Brown and Company, 1985. Letters of application and resumes are explained and illustrated on pages 365 to 382.

Lewis, Adele. *How to Write Better Resumes*. Woodbury, New York: Barron's, 1977. This practical, how-to book contains more than a hundred examples of action-getting resumes and practical suggestions for participating in an interview. The examples cover a wide range of specializations. The book disagrees with some of the suggestions offered in Unit 22. It, therefore, provides students with additional information for evaluating differing points of view concerning resume preparation.

Markel, Michael H. *Technical Writing: Situations and Strategies*. New York: St. Martin's Press, Inc., 1984. A good-news letter is illustrated on page 447. A negative letter is illustrated on page 448. Pages 461 to 485 give detailed information and examples related to the job-application procedure.

Roundy, Nancy, with David Mair. *Strategies for Technical Communication*. Boston: Little, Brown and Company, 1985. See Chapter 10: "Technical Letters and Memorandums," and Chapter 11: "Resumes and Letters of Application."

Swindle, Robert E. *The Concise Business Correspondence Style Guide*. Englewood Cliffs, N.J.: Prentice-Hall, Inc., 1983. Routine, good-news, bad-news, and persuasive letters, respectively, are illustrated and explained on pages 153 to 185. Word-processing methods are discussed on pages 138 to 150.

# Effective Use of Language in Technical Writing

# Correct Grammar

## OBJECTIVES

After reading this unit, you should be able to do the following tasks:

- Avoid sexism in writing.
- Apply the guidelines for grammar usage in written reports.
- Use the correct punctuation in writing.

## OVERVIEW

Why is the study of English grammar important in a technical-writing course? Grammar is the study of words, especially their forms and functions in a sentence. If words are the basis for all writing, effective writing must begin with correct grammar, including mechanics such as capitalization, punctuation, and spelling. In addition, grammatical errors, misplaced punctuation, and misspelled words draw attention to themselves, forcing the reader to focus thought upon them rather than upon the ideas given in the report. Interruptions of this kind, even occasional ones, often frustrate a reader and interfere with the concentration a reader needs to understand a message. Thus, correct grammar is essential for effective communication. Correct grammar also contributes to the authoritative tone of a message. By doing so, it increases the writer's prestige and probably his or her earning potential.

The student of technical writing already has a working knowledge of language and grammar rules and their correct use. Accordingly, the discussion in this unit is limited to a review of language and grammar. The unit first discusses nonsexist language; it then considers the guidelines of usage and punctuation that frequently confuse writers.

# NONSEXIST USAGE

*Sexism* in writing is the use of language that stereotypes, on the basis of sex, the characteristics and/or roles of men and women. There has been an ongoing movement toward eliminating such language in technical writing. Some of the following suggestions may help writers avoid sexist writing. For additional information on this subject, send for the National Council of Teachers of English publication titled *Guidelines for Nonsexist Use of Language* listed in the Suggested Readings at the end of this section.

## Avoid Excluding Women in Writing

It is no longer considered correct to use the masculine pronoun he (him, his, himself) to refer to any singular person whose gender is not known. Such expressions as "The manager . . . he," "The writer . . . he," and "The worker . . . he" seem to stereotype these roles as masculine. Guidelines for alternatives to the use of masculine pronouns follow.

- Change the subject from singular to plural.

| EXAMPLE | —*Sexist:* A restaurant manager orders his supplies weekly. |
|---------|-------------------------------------------------------------|
|         | —*Nonsexist:* Restaurant managers order their supplies weekly. |

- Reword the entire sentence so that a pronoun is not necessary.

| EXAMPLE | —*Sexist:* The carpenter can help make his community attractive. |
|---------|------------------------------------------------------------------|
|         | —*Nonsexist:* The carpenter can help to make an attractive community. |

- Replace the masculine pronoun with *one,* or *he* or *she (his or her).*

| EXAMPLE | —*Sexist:* Each student should have his book in class tomorrow. |
|---------|-----------------------------------------------------------------|
|         | —*Nonsexist:* Each student should have his or her book in class tomorrow. |

## Avoid Stereotyping in Job Titles

The titles of many occupations infer that workers are either all men or all women. Such titles should be changed to eliminate this inference.

<table>
<tr><td rowspan="3">**EXAMPLES**</td><td>*Sexist:* mailman</td><td>*Nonsexist:* postal carrier</td></tr>
<tr><td>*Sexist:* draftsman</td><td>*Nonsexist:* drafter</td></tr>
<tr><td>*Sexist:* stewardess</td><td>*Nonsexist:* flight attendant</td></tr>
</table>

# GRAMMAR USAGE

The items that most frequently cause errors in grammar are collective nouns used as subjects, parenthetical ideas that separate the subject and verb, indefinite pronouns used as antecedents, vague antecedents, and the correct us of *who, whom,* and *whose.* One grammar rule, the correct use of temporary hyphens (p. 292), is especially important in technical communication. Failure to apply it can cause serious misinterpretations of data.

## Collective Nouns as Subjects

Collective nouns include such words as *company, management, organization, class,* and *group.* As a subject, these words and other collective nouns are usually singular and are followed by a singular verb. Any pronoun referring to singular collective nouns must also be singular. Dictionaries are useful in identifying collective nouns and showing whether they are singular or plural.

**EXAMPLE**

**The company *has* made plans for technical training for *its* employees.**

Notice in this sentence that a singular verb, *has,* and a singular pronoun, *its,* follow the collective noun, *company.*

If the writer prefers using a plural verb and pronoun — and sometimes they seem logical — the sentence may be changed to read "The officers of the company *have* made plans for technical training for *their* employees." *Their* is correct because the subject is the plural noun *officers.*

## Parenthetical Phrases Between the Subject and Verb

When a subject is separated from its verb by a parenthetical phrase, only the subject determines the verb.

**EXAMPLE**    The vacuum tube, not the resistor or the condenser, is defective.

The subject is *tube*, a singular noun; therefore, the verb is singular.

**EXAMPLE**    Voltage, as well as current and resistance, was illustrated.

The only subject is *voltage*, a singular noun; the verb is singular. If the sentence is rewritten to read "Voltage, current, and resistance were illustrated," the compound subject — *voltage, current,* and, *resistance* — requires a plural verb.

## Indefinite Pronouns as Antecedents

The following indefinite pronouns are always singular: *anybody, anyone, anything, everybody, everyone, everything, nobody, no one, nothing, somebody, someone, something.* Any pronoun referring to one of these words must also be singular. Be sure not to only use the traditional masculine pronouns, which are considered sexist.

**EXAMPLES**    —Everybody has *his* or *her* book.
—Anyone *who* wants help must make an appointment.

Although *who* in the last example may be a singular or plural form, it must be considered singular in this sentence to agree with *anyone;* therefore, the verb following *who* must also be singular.

Indefinite pronouns are vague words because they make no specific reference to definite nouns. Also, most readers tend to incorrectly think of plurals when seeing such words as *everyone.* Therefore, the singular pronouns that refer to indefinite pronouns often sound incorrect. For these reasons, a good writer attempts to substitute specific nouns, as shown in the following examples.

**EXAMPLE**    Each (every) student (employee or other appropriate noun) has his or her book.

This statement can also be written in the plural: All students (employees or other appropriate plural noun) have their books.

---

**EXAMPLES**

—**One of you (us or the group) must express an opinion.**

—**Any person (employee or other appropriate noun) who wants to help must make an appointment.**

---

## Vague Antecedents

Vague antecedents occur whenever the reader fails to recognize immediately the word or words that such pronouns as *it, that, this, these,* and *those* refer to. Use the following three guides to avoid vague antecedents.

- Pronouns should not be used as beginning words in a paragraph.

---

**EXAMPLE**

**This indicates that the condenser is bad.**

---

What does *this* refer to? This leakage, perhaps; but some meaningful noun must follow the pronoun *this.*

- No noun that can be mistaken for an antecedent should come between the true antecedent and the pronoun.

---

**EXAMPLE**

**Because the carburetor in our car is defective, a new one must be purchased.**

---

This statement may make the reader question whether the carburetor or the car needs to be purchased. The following revision eliminates doubt.

---

**EXAMPLE**

**The defective carburetor in the car must be replaced.**

---

- The writer should not hesitate to repeat a noun when the repetition is necessary to the meaning of the sentence or the paragraph.

---

**EXAMPLE**

**A firm must pay its account within thirty days. If it fails to do so, it will be charged a late fee.**

---

The meaning is much clearer if the second sentence is rewritten as shown in the next example.

---

**EXAMPLE**

**Any firm failing to do so will be charged a late fee.**

---

Meaningful repetition is discussed in detail in Unit 29.

## Who, Whom

Choosing between *who* and *whom* and the compound forms *whoever* and *whomever* is sometimes a problem in written and spoken communications. These words are seldom used in formal reports. They are, however, used in interoffice reports, oral reports, and business letters.

When deciding whether to use *who* and *whoever* (nominative forms) or *whom* and *whomever* (objective forms), determine how the pronouns are used within their own clause according to the following guidelines.

- *Who* and *whoever* are the forms used for subjects and predicate nominatives.

---

**EXAMPLE**

**Please let us know who will be responsible for the test.**

---

*Who* is the subject of the verb *will be.*

---

**EXAMPLE**

**Who do you think should be sent on the field trip?**

---

*Who* is the subject of *should be sent.*

Occasionally, *who* is the subject of a clause that functions as the object of a preposition.

**EXAMPLE**

**We want to give the assignment to *whoever is best qualified*.**

Although the entire clause *whoever is best qualified* is the object of the preposition, *whoever* is correct because it is the subject of the clause.

*Who* or *whoever* is rarely used as a predicate nominative. One example is shown next.

**EXAMPLE**

**We have not decided *who* the new employee will be.**

- *Whom* and *whomever* are the forms used for direct objects and objects of prepositions in clauses.

**EXAMPLE**

**Patricia Jones is one of the supervisors *whom* we selected.**

*Whom* is the direct object of the verb *selected.*

**EXAMPLE**

**H. R. Smith is the person for *whom* this report is being written.**

*Whom* is the object of the preposition *for. (This report is being written* is an adjective clause modifying *whom.)*

- Careful writers can frequently avoid the use of *who* and *whom* by eliminating unnecessary words.

**EXAMPLES**

—**We selected Patricia Jones, one of the supervisors.**
—**The report is being prepared for H. R. Smith.**

- The general rule concerning who and *whom* has one exception: The subject of an infinitive is always in the objective case.

| | |
|---|---|
| **EXAMPLE** | ***Whom** do you want to write this report?* |

Again, awkward but grammatically correct sentences can often be effectively reworded.

| | |
|---|---|
| **EXAMPLE** | ***Who** should write this report?* |

# PUNCTUATION

Punctuation marks (commas, semicolons, colons, periods, hyphens, dashes, or parentheses) in or after a sentence indicate a shift in thought. For this reason, choosing the most appropriate punctuation requires judgment as well as knowledge of rules. The writer must decide what he or she is saying and then punctuate in such a way that the reader does not misinterpret the writer's meaning. General rules of punctuation are taught in elementary grammar-usage courses. A few rules often misinterpreted or violated in report writing are explained in this section.

## Commas Between Items in a Series

Some writers today omit the comma before the *and* when three or more words, phrases, or clauses are written in a series. However, technical writers are encouraged to use the comma. It is never considered incorrect and frequently avoids confusion, especially when the items in the series include the word *and* as part of one item.

The following example illustrates words in a series.

| | |
|---|---|
| **EXAMPLE** | *The effects of electricity can be seen in such appliances as clocks, radios, coffee pots, and toasters.* |

The next example illustrates phrases in a series. The comma after *communities* separates an introductory phrase from the main clause.

| | |
|---|---|
| **EXAMPLE** | *In our homes, in our offices, and in our communities, electricity is constantly working.* |

The following example illustrates clauses in a series.

**EXAMPLE**

**One can see that electricity is useful, that it is powerful, and that it can be dangerous.**

The next example shows items in a series that include the word *and*, a construction that can cause confusion if the final comma is omitted.

**EXAMPLE**

**Three types of gear trains are frequently used in automobile drivetrains. They are rack and pinion, ring and sun, and spur.**

## Commas and Semicolons Used for Punctuation of Clauses

Using the following guidelines helps to eliminate incorrect punctuation in a sentence that contains two or more clauses.

- When two independent clauses are separated by a conjunction *(and, but, or, nor, yet,* or *so)*, a comma is usually placed before the conjunction.

**EXAMPLE**

**Water freezes at 32 degrees on the Fahrenheit scale, but it freezes at 0 degrees on the Celsius scale.**

*Exception:* If each clause is short and is easy to understand, the comma may be omitted.

**EXAMPLE**

**The filter capacitors were replaced but the output voltage remained low.**

- When commas are used within a clause, a semicolon is used to separate two independent clauses.

**EXAMPLE**

**Many metals are heavier than water; but some, including potassium, sodium, and lithium, float on water.**

- When an adverbial clause precedes a main clause, a comma follows the adverbial clause.

**EXAMPLE**

**If an N pole and an S pole of two magnets are brought almost together, they attract each other.**

When the adverbial clause *follows* the main clause, no comma is used.

**EXAMPLE**

**Two magnets attract each other when the N pole and the S pole are brought almost together.**

- A semicolon is used between two closely related independent clauses that are not joined by a conjunction.

**EXAMPLE**

**A flashing is formed from sheet metal; it is used to waterproof joints or edges of a roof.**

Note:   The clauses are used as a compound sentence only when they are closely related. Otherwise, they are written as two sentences.

- When two clauses are separated by words functioning as conjunctive adverbs, such as *however, moreover, consequently, therefore, likewise,* and *nevertheless,* a semicolon precedes the conjunctive adverb and a comma follows it. (These words do not always function as conjunctive adverbs.) If, however, a short conjunctive adverb such as *then* or *hence* is used, the comma may be omitted. In the following example, a comma follows *however.*

**EXAMPLE**

**Overtime pay is not the answer to a long-term problem; however, it may be used for short-term emergencies.**

In the next example, the comma is omitted after *then.*

**EXAMPLE**

**To determine the lateral area of a prism, find the area of each side surface; then add these areas together.**

- When two dependent clauses are joined by a conjunction, no punctuation is used between them.

<table>
<tr><td>

**EXAMPLE**
</td><td>

**If an object is not round and if the object can change shape easily, a pi tape is often used to determine the diameter.**
</td></tr>
</table>

The two clauses introduced by *if* are dependent clauses.


## Colons

A colon signals to a reader that something important is about to be said. When used in general text, a colon follows only a complete statement.

<table>
<tr><td>

**EXAMPLE**
</td><td>

**The following are the three classes of electric circuits: series circuits, parallel circuits, and series-parallel circuits.**
</td></tr>
</table>

Notice that the list follows a complete statement; the list, therefore, is preceded by a colon.

In the next example, the list follows an incomplete statement; therefore, no punctuation precedes the list.

<table>
<tr><td>

**EXAMPLE**
</td><td>

**The three classes of electric circuits are series circuits, parallel circuits, and series-parallel circuits.**
</td></tr>
</table>


## Dashes

Dashes are attention-getting devices that emphasize the information written between them. Because the reader's attention should normally be focused upon the entire idea expressed in technical reports rather than upon one small part, commas should usually be substituted for dashes. However, when a list of items interrupts the main parts of a sentence, the entire list must be set off by dashes.

<table>
<tr><td>

**EXAMPLE**
</td><td>

**The three units in the magnetic circuit — flux, magnetomotive force, and reluctance — correspond to current, voltage, and resistance in the electric circuit.**
</td></tr>
</table>

# Hyphens

The hyphen, perhaps more accurately described as a mark of spelling than a mark of punctuation, creates problems for most writers. The problems develop because hyphens may be permanent (used each time a particular word is written) or may be temporary (used only under certain conditions).

**Permanent Hyphens**   The writer must rely upon the dictionary to determine whether some words, especially compound nouns such as *centerline, editor in chief,* or *jack-of-all-trades,* are written as one word, more than one word, or hyphenated. Seven general guidelines, however, solve most problems relating to permanent hyphens.

- Prefixes such as *non-, pre-, pro-, trans-,* and *un-* are separated from a proper noun by a hyphen.

**EXAMPLES**

| | |
|---|---|
| non-Christian | pre-Grecian |
| pro-African | trans-Canadian |
| un-American | |

- All words beginning with *pre-, pro-,* or *trans-* in which the second word is not capitalized are written as one word.

**EXAMPLES**

| | |
|---|---|
| predate | transamination |
| predetermine | transducer |
| prorate | transect |
| prothorax | translucent |

- Almost all words (excluding those related to the first rule) that begin with the prefixes *co, ex-, non-, re-, un-,* and *under-* are written without a hyphen.

**EXAMPLES**

| | |
|---|---|
| cooperate | nonvascular |
| coordinate | reemission |
| coeducation | reevaluate |
| exanimate | refilter |
| exponent | refilm |
| nonmechanical | unabsorbable |
| nonparallel | unacclimated |

Almost all dictionaries show long, complete lists of this kind of word. Exceptions to this rule include a very few words that could be misunderstood or mispronounced.

**EXAMPLES**

| | |
|---|---|
| **co-op** | **re-cover (to cover again)** |
| **co-opt** | **re-claim (to demand the return of)** |
| **re-creation (to create anew)** | |

- Hyphens are used in words that begin with the prefix *self-*.

**EXAMPLES**

| | |
|---|---|
| **self-energizing** | **self-perpetuating** |
| **self-exciting** | **self-starter** |

*Selfish, selfless,* and *selfsame* are written as one word because *self* is not a prefix in these words.

- Hyphens are used in words beginning with *ex-* when *ex-* means "previous."

**EXAMPLES**

| | |
|---|---|
| **ex-mayor** | **ex-president** |

- Some words that begin with *well* as a prefix are hyphenated; others are not.

**EXAMPLES**

| | |
|---|---|
| **well-advised** | **adjective** |
| **wellborn** | **adjective** |
| **well-meaning** | **adjective** |
| **well drain** | **noun** |
| **wellhead** | **noun** |
| **well-wisher** | **noun** |
| **well-drain** | **verb** |

These examples show that no obvious rule applies to words beginning with *well*. A dictionary should be checked for correct forms. For words not listed in the dictionary, writers apply rules for temporary hyphens.

- A word that is written as one word or as a hyphenated word when used as a noun is written as two words when used as a verb-adverb combination.

**EXAMPLE**

**The *breakdown* of the motor resulted from defective materials.**

*Breakdown* is a noun.

**EXAMPLE**

**Did the motor break down because of defective materials?**

*Break* is a verb; *down* is an adverb.

**EXAMPLE**

**The *tune-up* of the engine required 2 hours.**

*Tune-up* is a hyphenated noun.

**EXAMPLE**

**He can *tune up* the engine in 5 hours.**

*Tune* is a verb; *up* is an adverb.

The question of why *breakdown* is one word and *tune-up* is hyphenated cannot be answered. Compound words similar to these must always be checked in the dictionary.

**Temporary Hyphens**   Temporary hyphens are usually found in two or more adjectives or words that are functioning as adjectives. These hyphens have often been ignored, even in some professional writing. However, their correct use is essential to a reader's immediate awareness of a writer's intended meaning. The following guidelines clarify the use of temporary hyphens.

- When two or more words used as adjectives function as a single modifier and precede the noun they modify, one or more hyphens join these modifiers into one word. (This kind of adjective is used continually in technical reports.)

**EXAMPLES**

—The *high-voltage* lines are dangerous.
—An *up-to-the-minute* report was expected.
—*Direct-current* series motors are *variable-speed* motors.

- When any words similar to the hyphenated words illustrated in the first rule do not precede a noun, they are not hyphenated.

**EXAMPLES**

—Measuring high voltage is dangerous.
—Many motors operate   on direct current.
—The notes read this morning were up to the minute.

In these examples, *voltage, current,* and *minute* no longer function as part of a compound adjective; they function as nouns.

Exception: *Up-to-date* is always hyphenated.

**EXAMPLES**

—He wrote an  *up-to-date* progress report.
—The progress report was *up-to-date.*

- A hyphen is used between an adverb and a present or past participle that precedes a noun unless the adverb ends in *-ly.*

**EXAMPLE**

The department had a *fast-moving* production line.

*Fast* does not end in *-ly.*

**EXAMPLE**

The *recently built* bridge already needs to be repaired.

*Recently* end in *-ly.*

- When the first word of a compound modifier preceding a noun is written in the comparative or superlative degree, a hyphen is used.

**EXAMPLE**

**Better-made radios is the company's goal.**

*Better* is comparative.

**EXAMPLE**

**The fastest-selling commodities are not necessarily the best.**

*Fastest* is superlative.

- A suspended hyphen is used when a series of compound adjectives precedes the noun, but the second word in the compound adjective is stated only at the end of the series.

**EXAMPLE**

**The work required the use of 1-, 3-, and 4-foot tapes.**

When the sentence is written as "The tapes required for the work must be 1, 3, and 4 feet in length," no hyphens are used.

- Fractions written as an adjective-noun combination are not hyphenated. Fractions written as adjectives are hyphenated because they are compound adjectives.

**EXAMPLE**

**Three fourths of the report has been completed.**

The subject of this sentence is *fourths; three* is a single adjective modifying *fourths.*

━━━━━━━━━━━━━
**EXAMPLE**
━━━━━━━━━━━━━
**Quarks are assumed to have either one-third or two-thirds the electric charge of electrons and protons.**

*One-third* and *two-thirds* are used as single words modifying *charge;* therefore, they are hyphenated. Most fractions used in technical writing are adjective-noun combinations, though, and are not hyphenated.

━━━━━━━━━━━━━
**EXAMPLE**
━━━━━━━━━━━━━
**Computers assist in interpreting almost one fifth of the electrocardiograms made in the United States.**

*Fifth* is a noun; *one* modifies it.

■ Compound numbers between *twenty-one* and *ninety-nine* are hyphenated.

━━━━━━━━━━━━━
**EXAMPLE**
━━━━━━━━━━━━━
**Thirty-one of the men work in the electronics department.**

Temporary hyphens are also used to divide words at the end of a typed line. Rules for using the hyphen for word division are discussed in Unit 7.

# SUMMARY

1. Correct grammar is the foundation for standard English, a basic principle in technical writing.

2. This unit explains only rules of grammar that have proved to be unfamiliar to many technical writers. Students who need general help in grammar should purchase and use a grammar textbook.

3. Temporary hyphens are important. Their correct use is essential to clarity in written communication.

4. A dictionary is a necessary tool for every technical writer.

5. Using correct grammar as an aid to effective communication involves not only a knowledge of grammar rules but also judgment and an awareness of the reader's needs.

UNIT **24**

# SUGGESTED ACTIVITIES

**A.** The following sentences may be grammatically correct or may contain one or more errors in grammar and punctuation. If the sentence is correct, write C at the end of it. If it is incorrect, correct the errors; then, using complete sentences, state a guideline that justifies the correction. If necessary, rewrite the sentence.

1. After the parts are assembled, but before the device is checked, the ohm-meter must be calibrated.

2. The supervisor offered a bonus to whoever was willing to work overtime.

3. Increasing the input voltage by means of an autotransformer is a commonly-used method for increasing the output-voltage of a power supply.

4. The company was proud of their accomplishments because they had increased production of capacitors, resistors and vacuum tubes during 1980.

5. Anyone who works in a research department learns that they must cooperate with others.

6. Because only one electron is in the outer shell of a hydrogen atom, it is said to have a valence of 1.

7. The diode, triode, and tetrode are two electrode, three electrode and four electrode electron tubes respectively.

8. One-oxygen atom, when combined with two-hydrogen atoms, create a molecule of water.

9. Impurities existing in the transistor housing or on the surface of the transistor can diffuse into the body of the transistor and change the internal structure, however, at normal temperatures, it is quite small and can usually be ignored.

10. The radio is built to operate on alternating current power or direct current power.

11. Whom will Helen be working for?

12. The need for unifying different road and street systems were obvious.

13. An effective traffic control device should meet five basic requirements, (1) fulfill a need, (2) command attention, (3) convey a clear, simple meaning, (4) command the respect of road users, and (5) give adequate time for proper driver response.

14. Every one of the tests were conducted under controlled conditions.

15. Has the supervisor decided who she wants to wire these components?

**B.** Select at least ten guidelines from this unit. Choose those that are unfamiliar or that are especially difficult to remember. Using subject matter from other classes, write a sentence illustrating each guideline chosen. Underline the part of the sentence that applies to the guideline; then, state the guideline. All sentences must be grammatically correct; spelling and punctuation must also be correct. Use the objective style of writing explained in Unit 1, and make the sentence meaningful to the reader.

C. Check the dictionary to determine whether the following words should be hyphenated, written as two or more words, or written as one word.

1. breast bone
2. breech loader
3. compass card
4. complementary angles
5. cost plus
6. counter clock wise
7. counter irritant
8. counter shaft
9. electro magnetic
10. foot pound
11. give and take (noun)
12. gram molecular
13. hydro airplane
14. hydro carbon
15. hydroxyl amine
16. idler wheel
17. journey work
18. kilogram meter
19. kilo watt hour
20. never the less

D. Bring to the classroom any informative writing that may contain grammatical errors. The writing sample may be from business letters, reports, or other messages obtained from industry, social service organizations, or other classroom courses. Discuss the errors and, if necessary, change sentences that seem awkward or vague.

# Effective Sentence Structure

After reading this unit, you should be able to do the following tasks:

- Use practical sentence lengths in written reports.
- Determine and use effective word order in sentences.
- Recognize and correct construction errors in sentences.

## OVERVIEW

A well-written sentence is always grammatically correct, but a grammatically correct sentence is not always well written. Sentence structure, like grammar, helps determine the quality of a report or letter. The *structure* of a sentence — the way a sentence is developed — is the arrangement of words, phrases, and clauses. Effective, interesting reports result from well-planned sentences in which the structure produces conciseness, completeness, and clarity. Three major topics that help in the construction of a well-written sentence are discussed in this unit: sentence length, word order in sentences, and construction errors.

# SENTENCE LENGTH

In technical writing, sentences have only one purpose: to transmit information to the reader. Short, organized sentences accomplish this purpose. Long, involved sentences are often written to impress, not to inform, the reader.

At least 80 percent of all sentences used in technical reports should range between ten and twenty words in length. Sentence length should vary, however, so this word range should serve only as a guideline, not as a rigid standard.

# WORD ORDER IN SENTENCES

Word order — the arrangement of words in a sentence — creates a variety of effects. By changing words from one position to another, a writer can build suspense, arouse emotion, or transmit information. Of these, technical writers are concerned only with transmitting information. They achieve this goal by understanding the importance of natural word order, the uses for active and passive voice, and the problems associated with weak introductory words and telescopic sentences.

## Natural Order

Three kinds of word arrangement may be used in the construction of English sentences:

1. Natural order, in which the subject precedes the predicate
2. Interrupted order, in which the subject is placed between two parts of the predicate
3. Inverted order, in which the subject follows the predicate

Most sentences in technical reports should be written in natural order, a word arrangement that presents information logically and clearly. The sentence in the following example is written in natural order. The meaning builds systematically as the reader progresses through the sentence.

**EXAMPLE**

**The film was placed in the film holder.**

Inverted order, such as "In the film holder was placed the film," withholds the meaning until the subject is read at the end of the sentence. Because inverted order builds suspense rather than meaning, it has little value to technical writers and should be avoided.

Natural order is recommended for most sentences in technical writing. If not used skillfully, however, it develops into a monotonous rhythm, focusing the reader's attention more upon style than upon meaning. See the next example.

**EXAMPLE**

A sulfur dioxide analyzer system provides continuous analysis of combustion emissions. It combines the use of a photometric analyzer with integrated sampling systems. Such a device provides the means of developing a method for reducing air pollution economically. The analyzer system measures the amount of polluting emission.

The sentences in this example, aside from giving misleading information, have three major sentence weaknesses:

1. They have almost identical sentence structure.
2. Multisyllabic words occur too frequently.
3. The sentences vary only slightly from one another in length.

A simpler, more readily understood version of the example follows.

**EXAMPLE**

A sulfur dioxide analyzer has two major parts: a photometer and a sampling system. The photometer measures the density of sulfur dioxide gas sent into the analyzer from the atmosphere. At the same time, the sampling system continually monitors and gives a time-averaged reading of this gas. The readout provides information needed to control air pollution from sulfur dioxide.

The following guidelines help prevent weaknesses in sentence structure.

- Vary sentence length. More than two sentences that are almost equal in length should rarely be written consecutively.

- Vary word length. Many words containing three or more syllables should not follow one another.

- Use transitional devices — words, phrases, and clauses that not only change the tone of a sentence but also lead the reader meaningfully from one sentence to the next.

- Vary word order. Occasionally, use interrupted order in which an adverbial phrase or clause is shifted from its normal position in the predicate to the beginning of a sentence.

| | |
|---|---|
| **EXAMPLE** | **On a 132-column printer, up to ten columns of tables can be created.** |

The adverbial phrase at the beginning of this sentence creates interrupted order. Here is another example.

| | |
|---|---|
| **EXAMPLE** | **Because semiconductor material is affected by heat, the characteristics of a P–N diode will change as its temperature changes.** |

The natural order of this sentence has been changed by introducing it with an adverbial clause.

Occasionally, a one-word adverb may introduce a sentence.

| | |
|---|---|
| **EXAMPLE** | **Sometimes, a motor will not start because the motor switch is defective.** |

## Active and Passive Voice

When the active voice of the verb is used, the subject acts.

| | |
|---|---|
| **EXAMPLE** | **The drafter improved upon the line quality.** |

When the passive voice is used, the subject is being acted upon.

| | |
|---|---|
| **EXAMPLE** | **The line quality was improved upon by the drafter.** |

The passive voice creates a passive tone and is, therefore, less dynamic than the active voice. However, when the active voice is used, the emphasis is placed on the subject of the sentence, which is usually the person engaged in the research, process, or analysis discussed in the report. Therefore, the doer rather than the thing being done occupies the first or primary position in the sentence. Controversies about whether active or passive voice should be used in reports have continued for many years. No rule applicable to all technical writing has been established.

The decision about voice in technical reports centers on the question "Should a report emphasize the person performing the action, or should it emphasize the subject about which the report is being written?" Even though the answer to the question seems obvious, a report is rarely written entirely in the passive voice. For example, one student described the procedure used in producing a newspaper advertisement from a rough copy. The markup person was important; the markup person's duties were also important. The student effectively combined active and passive voice and, by doing so, stressed both the doer and the thing being done: "A markup person must select the typefaces to be set; in most ads, only one typeface is used because wide variety can be obtained by the use of light, medium, bold, heavy, italic, or condensed type."

Writers who are aware of what is important in a specific report do not slip carelessly from one voice to another, nor do they let an obsession with rules concerning active and passive voice stifle the logical development of the subject.

## Weak Introductory Words

The expletives *there* and *it* are weak sentence openers because they have no meaning and no grammatical value. They are introductory and nothing more. The frequent use of these words suggests that the writer is afraid to make a strong statement about his or her information. *There* and *it* also force a sentence into inverted order, which, as stated previously, should be avoided in technical writing.

Consider the following examples.

| EXAMPLE | **There is a vacuum hose connected to the distributor.** |
|---|---|

This sentence is shorter and more forceful when written as "A vacuum hose is connected to the distributor."

| EXAMPLE | **There are six major pieces of equipment needed for this experiment.** |
|---|---|

If this sentence is changed to "Six major pieces of equipment are used for this experiment," the message is conveyed more quickly and easily.

---
**EXAMPLE**

**It should be noted that the procedure used in the test for strength may be repeated several times.**

---

This sentence is unnecessarily long, wordy, and weak. It is improved when written as "The procedure used in the test for strength may be repeated several times."

## Telescopic Sentences

In telescopic sentences, small but important words like *a*, *an*, and *the* are omitted. Some students, in an effort to be concise, develop telescopic style. Others believe that this brevity gives a tone of dignity and authority to writing. "Sent message to office requesting report" is brief, but it is also rude, incomplete, and ambiguous.

Telescopic writing in reports creates one or more of three problems:

1. It may force the reader to read so rapidly that meaning is lost.
2. It may force the reader to subconsciously supply the missing *a*'s, *an*'s or *the*'s.
3. It may force the reader to read hesitantly, sensing that something is wrong with construction or meaning.

The following example of a paragraph made up entirely of telescopic sentences illustrates these three problems.

---
**EXAMPLE**

**Temperature is determined by level of mercury contained in thermometer tube. If temperature surrounding tube is hot, mercury expands and rises up to temperature around it. As temperature drops, so does mercury in tube.**

---

When writers understand the disadvantages of telescopic writing, they do not feel a sense of wordiness when they write full sentences that include all the small words needed for meaning and readability.

# CONSTRUCTION ERRORS

Construction errors are usually considered to be errors in grammar. However, they are introduced in this unit rather than in Unit 24 because rules relating to them are more easily understood after sentences in general have been defined. The construction errors that must be recognized and avoided in all technical writing include incomplete sentences, dangling verbals, faulty parallelism, incomplete comparisons, and misplaced modifiers.

## Incomplete Sentences

Using incomplete sentences, those that do not contain a complete subject and predicate and do not express a complete idea, is inexcusable in technical communication. Incomplete sentences in reports usually result from carelessness, not from lack of knowledge. They occur when a writer punctuates incorrectly; fails to organize the information; or mistakes a gerund, participle, or an infinitive for a main verb. When one student wrote "The solution was allowed to cool slightly. For only about 5 minutes," a phrase was used as a complete sentence. This error could have been avoided if unnecessary words had been omitted and if the information had been presented specifically as in "The solution was allowed to cool for 5 minutes."

Incomplete sentences similar to the following example occur when a writer fails to check and organize the list of materials before writing a report.

**EXAMPLE**    **The materials needed for a minor tune-up are eight spark plugs that have the correct heat range, one set of breaker points, and one condenser. Also a tube of distributor-cam lubricant.**

The sentence should have been written "The materials needed for a minor tune-up are eight spark plugs that have the correct heat range, one set of breaker points, one condenser, and one tube of distributor-cam lubricant."

Incomplete sentences are avoided when the writer recognizes a complete sentence, organizes the material, and checks the first draft of the report for errors in sentence construction.

## Dangling Verbals

Verbals are verb forms that, with rare exceptions, do not function as verbs in sentences but as nouns, adjectives, or adverbs. Verbals are divided into three kinds: gerunds, infinitives, and participles.

When a verbal is used as an adjective or adverb or as part of an adjective or adverb, it may *dangle* (not logically modify any part of the sentence) unless a noun or pronoun performing the action expressed in the verbal is included in the sentence.

**EXAMPLE**    **Having selected the site for the pier and grade-beam system, unwanted plants, loose soil, and other debris were removed.**

In this sentence, the writer has failed to say who selected the site. Therefore, *having selected* is a dangling verbal. Dangling verbals are unacceptable in technical writing because they can mislead or confuse the reader and because they are errors in the correct use of grammar.

**Dangling Gerunds**  A gerund is a present participle (-*ing*) form of a verb that substitutes for a noun in a sentence. It, therefore, can function as a subject, a direct object, a predicate nominative, an object of a preposition, or an appositive. Only gerunds used as objects of prepositions can dangle. Here are some examples.

**EXAMPLES**

—After thoroughly mixing the ingredients, they were poured into the mold.

—By connecting a spring to the electromagnet, the pointer is forced toward 0 when no current is flowing in the electromagnet.

—By using freshly drawn blood, a more nearly accurate analysis can be made.

Sentences like these tend to make a reader hesitate and perhaps try to rephrase them. Thus, they usually interrupt the logical development of information in a written communication. The following guidelines help technical writers avoid these and other dangling gerunds.

- Make the complete sentence active. (Clarify who is performing the action.)

**EXAMPLE**

After thoroughly mixing the ingredients, the technician poured them into the mold.

- Change the prepositional phrase containing the gerund to an adverb clause.

**EXAMPLE**

When a spring is connected to the electromagnet and no current is flowing in the electromagnet, the pointer is forced toward 0.

- Construct a simpler sentence.

**EXAMPLE**

Freshly drawn blood gives (or permits) a more nearly accurate analysis.

Some gerunds that grammatically can be classed as danglers may not confuse the reader and are, therefore, acceptable. For example, in "Plastic plates are part of a new process used in printing newspapers" *printing* is acceptable. The reader is not likely to question who does the printing. Thus, effective communication often depends upon a writer's ability to combine judgment with rules.

**Dangling Infinitives**  Infinitives are made up of the preposition *to* followed by a verb form. Infinitives may substitute for nouns, adjectives, or adverbs in a sentence. (Occasionally, they are also used as verbs in a dependent clause.) Infinitives may be active or passive.

**EXAMPLES**

| | |
|---|---|
| to see | present active |
| to have seen | perfect active |
| to be seen | present passive |
| to have been seen | perfect passive |

A present active infinite used as an adverb occasionally dangles.

**EXAMPLE**

To determine the power output of a transmitter, the voltage and current in the final stage must be known.

This sentence is improved if the infinitive phrase is converted to a clause.

**EXAMPLE**

The power output of a transmitter can be determined if the voltage and current in the final stage are known.

The following similarly constructed sentence is not likely to distract the reader and is acceptable.

**EXAMPLE**

To determine the number of cars the lot would accommodate, a survey was made.

In the following sentence, a reader cannot be sure who does the explaining; a revision is needed.

**EXAMPLE**

The supervisor called the crew together to explain the cause of the power failure.

Here are two possible revisions.

**EXAMPLES**

—The supervisor called the crew together and explained the cause of the power failure.

—The supervisor called the members of the crew together and asked them to explain the cause of the power failure.

**Dangling Participles**  A participle is a verb form that functions in a sentence as an adjective and thus must clearly relate to an expressed noun or pronoun. Any one of five verb forms may be used as a participle.

**EXAMPLES**

| | |
|---|---|
| making | present active participle |
| being made | present passive participle |
| made | past passive participle |
| having made | perfect active participle |
| having been made | perfect passive participle |

Any form of a participle can be used incorrectly. It can dangle, wink, or trail. In the following sentence, *having completed* dangles; the person who completed the project is not named.

**EXAMPLE**

Having completed the project, the information concerning it was given to the foreman.

The sentence may be correctly written in one of two ways.

**EXAMPLES**

—Having completed the project, the drafter gave the information concerning it to the foreman.

—After the project was completed, the information was given to the foreman.

In the next sentence, the participle *pinched* winks — that is, can modify two things. It can modify either the technician or the skin.

**EXAMPLE**

When pinched between the thumb and forefinger, the medical technician noticed that the skin was dry and shrunken and remained inelastic.

The sentence can be corrected in various ways. Here is one possible revision.

**EXAMPLE**

**The medical technician noticed that the skin, when pinched between the technician's thumb and forefinger, was dry, shrunken, and inelastic.**

In the following sentence, *resulting* is a trailing participle. Because it seems to modify a clause instead of a noun or pronoun, it should be avoided.

**EXAMPLE**

**The motor was tuned up *resulting* in a smoother start.**

One possible revision is, "A smoother start resulted from the motor tune-up."

## Faulty Parallelism

The term *parallel* means that two or more things are equal or similar in all essential details. This definition can be applied to parallelism in English. Two or more words, phrases, or clauses that are equally important should have balanced, equal construction in a sentence. Failure to observe this rule causes faulty parallelism — an error in sentence structure that causes awkwardness and confusion.

Faulty parallelism, which is the result of unorganized thinking or carelessness, occurs in words, phrases, or clauses written in a series or in itemized lists. See the following example.

**EXAMPLE**

**The tools required for this project include a wrench, a hammer, and a screwdriver is also needed.**

This sentence tells about three equally important tools. Two are merely named, but the third is placed in a separate clause. So this construction is faulty parallelism. The following example shows one possible revision that corrects the faulty parallelism.

**EXAMPLE**

**The tools required for this project include a wrench, a hammer, and a screwdriver.**

The parallelism is also faulty in the next example. A prepositional phrase is connected to a dependent clause.

**EXAMPLE**

**This report has described the reasons for minor tune-ups and what is involved in tuning up an engine.**

The faulty parallelism can be corrected in many ways. Here is one possibility.

**EXAMPLE**

**This report has explained why minor tune-ups are necessary and what is involved in tuning up an engine.**

Now, the two equal ideas are balanced because each has been placed in a dependent clause.

## Incomplete Comparison

Comparative forms such as *higher, more efficient,* or *more frequently* indicate that two items are being compared with each other or that one item is being compared with other similar items. A writer must, therefore, be certain that the comparison is complete and that the reader can easily see the full comparison. The following example was taken from a Star Wars report.

**EXAMPLE**

**Chemical lasers, whose beams are produced by chemical reaction, would need to be far more powerful.**

The lasers need to be far more powerful when, why, or for what? If a preceding sentence has not made the comparison complete, the sentence must be lengthened so that all parts of the comparison are included.

In the example, both why and when seem to be missing. Though the author never clarified his comparison, the surrounding text suggested the meaning expressed in the following example.

**EXAMPLE**

**Chemical lasers, whose beams are produced by chemical reaction, would need to be more powerful than they are in order to destroy warheads.**

Although conciseness is encouraged in report writing, clarity or completeness can never be sacrificed.

## Misplaced Modifiers

A careful writer avoids misplacing words and phrases. One rule of correct grammar usage specifies "Place a modifier near the word it modifies." This rule applies to all adjectives, adjective phrases, and adjective clauses. It should also apply to adverbs, adverbial phrases, and adverbial clauses. For some unexplainable reason, however, adverbs are more flexible than adjectives. They can appear at the beginning of a sentence, before the verb, between two parts of the verb, or after the verb without distorting meaning. Therefore, although English has a rule that supposedly solves the problem of misplaced modifiers, writers must use judgment to decide whether they have placed modifiers where they are most meaningful.

Consider the following example.

| EXAMPLE | She wrote a report to her supervisor explaining the operation of the machine. |
|---------|---|

This sentence says that the supervisor is doing the explaining. It must be rewritten so that the participial phrase is placed near the word it modifies. Here is a possible revision.

| EXAMPLE | The report explaining the operation of the machine was sent to the supervisor. |
|---------|---|

The position of such words as *almost, even, just, nearby,* and *only* can affect the meaning of a sentence. These words should, therefore, be placed as closely as possible to the word each modifies.

In the following example, two pairs of sentences are given. Notice the difference in meaning between the two sentences in each pair.

| EXAMPLES | —The door frames had just been installed. |
|----------|---|
|          | —Just the door frames had been installed. |
|          | —The illustrator has almost completed all the sketches. |
|          | —The illustrator has completed almost all the sketches. |

*Only* can have as many as four different positions that result in four different sentence meanings.

---

**EXAMPLES**

—This machine only punched 300 holes in an hour.

—This machine punched only 300 holes in an hour.

—This machine punched 300 holes in only an hour.

—Only this machine punched 300 holes in an hour.

# SUMMARY

1. Like basic grammar, sentence structure contributes significantly to the quality of a written communication.

2. Sentences must primarily transmit specific information but must also maintain reader interest so that the information is readily understood.

3. Short sentences are usually better than long ones; however, length and word arrangement must vary to prevent monotony.

4. A careful writer recognizes and usually avoids weak introductory words. He or she also recognizes and avoids telescopic writing, dangling verbals, incomplete sentences, incomplete comparisons, and misplaced modifiers.

5. The need for a writer to use judgment while developing sentences cannot be overemphasized. Judgment helps create individual but effective style; therefore, it is as important as — if not more important than — conformity to sentence-structure rules.

---

UNIT **25**

# SUGGESTED ACTIVITIES

A. Each of the following sentences has one or more construction errors. Rewrite the sentence correctly and objectively; then, indicate which of the listed errors was corrected.

   —Telescopic writing
   —Incomplete comparison

—Dangling gerund
—Dangling participle
—Dangling infinitive
—Misplaced modifier
—Faulty parallelism
—Unnecessary expletives

1. Mechanical failure of a transistor results from poorly made connections, excessive strain on sections of transistor, and when too much heat is applied during soldering operation.

2. Precautions against unsterile conditions while manufacturing transistors even extend to personnel.

3. After the motor was replaced, the machine ran more efficiently.

4. A pointer is placed on the electromagnet between two retaining pins.

5. In measuring resistance, many techniques are similar to those used in measuring voltage and current.

6. It is obvious that the voltage in this circuit is too high.

7. This department now has fewer complaints from management.

8. The tubes only needed to be tested once.

9. Using the oscilloscope, the peak-to-peak voltage was measured.

10. The furnace chamber was 13 feet long and 2 feet wide while the height was 3 feet.

11. There are two sides parallel to each other in a trapezoid.

12. Material is usually called an insulator and has low heat conductivity when the material has high resistance to heat flow.

13. It is apparent that heat can travel by three methods: conduction, convection, and also by radiation.

14. There are many kinds of metals that transfer heat through conduction in which the heat doesn't move at once from one part of the metal to all parts but gradually.

15. Tests indicated a ruptured vessel in the brain which caused death.

16. By comparing weekly readings on the inclinometer, the amount, depth, and direction of earth movement can be determined.

17. The Veterans Administration Hospitals have been accommodating an ever-increasing number of patients because of the Vietnam War and because legislation now permits medical care for families of these veterans at these hospitals.

18. The liquid was siphoned from the tank, leaving a residue of silt.

19. Remarkable adhesives are now available that transform, with the addition of catalysts and moderate pressure, from free-flowing liquids to rigid plastics.

20. Having placed 1 milliliter of whole blood in a 10- by 75-millimeter test tube, it was allowed to clot in a 37-degree Celsius water bath.

B. From the material taught in other classes, compose two informative sentences illustrating each of the following. Check the sentences to see that they express information concisely and directly.

—Correctly used gerunds functioning as objects of prepositions
—Correctly used participles
—Correctly used infinitives
—Parallel construction of items written in series
—Complete comparisons

Use correct grammar, punctuation, spelling, and sentence structure in all sentences. Write seven of the sentences in natural order and three of them in interrupted order.

C. Bring to the classroom any sentences from technical literature or textbooks that seem to be carelessly constructed. If possible, determine the cause of weak sentence structure; then, improve the sentence.

# Correct Use of Numbers

## OBJECTIVES

After reading this unit, you should be able to do the following tasks:

- Use the guidelines for exact and general numbers in a written report.
- Use the guidelines for fractions in a written report.

## OVERVIEW

During the past two decades, technical writers have had difficulty determining how numbers should be written in reports and other business communications. Firmly established rules for numbers, especially in technical writing, have not been clarified. Many of the rules vary from textbook to textbook and from company to company. Variations exist partly because America has moved rapidly toward increased industrialization and technology in which numbers, especially exact numbers, play an important role.

Perhaps the most important guideline a writer can follow when using numbers is to remember the reader and present numbers in a way that the reader readily understands. Another important consideration is company policy. A cooperative writer abides by company practices or, if justified, tries to change them. The guidelines given in this unit were selected because they are extensively used and create meaningful text.

Technical writing makes a distinction between *exact numbers* that need emphasis and *general numbers* that should not be considered more important than the other words in the sentence. Because numbers written as figures are more emphatic than numbers written as words, exact numbers, unless they begin the sentence, are always written as figures. General numbers are usually written as words. The following guidelines help clarify these two groups of numbers. Even in some of these guidelines, two alternatives are offered because both alternatives have been widely but not universally adopted.

# EXPRESSING NUMBERS

The following guidelines help technical writers determine when to use digits and when to use words to express numbers.

- Numbers that begin a sentence are always written as words. This guideline has no exceptions.

**EXAMPLES**

—Seventeen percent of the power was lost.

—Three thousand ohms of resistance were in the circuit.

—Twenty-five hundred dollars were used for training programs.

A number that cannot be expressed in fewer than four words should not be used as the first word of a sentence.

**EXAMPLES**

—The power loss was 17.9 percent.

—The company employed 365 people.

—The budget included $15,500 for equipment.

- Nonspecific numbers in written material are written as words in accordance with either of two guidelines.

  1. Numbers below 100 are written as words.

**EXAMPLES**

—The thirty-six punched cards are on the shelf.

—The 200 employees planned to strike.

  2. Numbers that can be expressed in one or two words are written as words.

**EXAMPLES**

—The manual contains one hundred illustrations of electronic devices.

—The book has 126 pages.

- Elements of time such as hours, days, weeks, and months are considered to be general terms. Numbers preceding them are written as words or figures in accordance with the previous guideline.

**EXAMPLES**

five hours

125 days

seven weeks

one hundred (or 100) years

An exception to this guideline must be considered for technical writing. Precise timing required for various processes is an exact number and is expressed in figures.

**EXAMPLE**

The film is developed for 12 minutes at 70 degrees Fahrenheit.

- If a sentence contains many numbers, especially in lists or in items in a series, one of two guidelines may be used:

  1. If the list is short and if all the numbers contain only one or two digits, the numbers may be expressed in either words or figures.

**EXAMPLE**

The eighteen electrons in argon are divided into three shells; two electrons are in the first shell, eight electrons are in the second shell, and eight electrons are in the third shell.

The numbers as written in this sentence are acceptable. Note, however, that the numbers are more easily read and remembered if figures are used.

**EXAMPLE**

The 18 electrons in argo are divided into three shells; 2 electrons are in the first shell, 8 electrons are in the second shell, and 8 electrons are in the third shell.

  2. If one or more numbers in the list contain three or more digits, figures must be used.

**EXAMPLE**

Please prepare 125 copies of the instructions, 10 copies of the blueprints, and 15 pictures of the diagram.

- When one number immediately follows another number, the first number is spelled out but the second number is written as a figure. In technical writing, the first number is usually a general one and the second is a specific one. This guideline, therefore, agrees with the recommendation that general numbers be written as words and that specific numbers be written as figures.

**EXAMPLE**

**They ordered five 16-foot boards; they also ordered twenty-five 8-foot boards.**

- Figures are always used when numbers express exact units of measure, distance, dimensions, cubic capacity, weights, amounts of money, and percent. (This rule is secondary to the first rule. Even exact numbers must be written as words if they begin a sentence.)

**EXAMPLES**

| | |
|---|---|
| **7 feet** | **15 cents** |
| **12 cubic yards** | **$82** |
| **(inches, feet)** | **27 percent** |
| **8 gallons** | **20 pounds** |

Note that the dollar sign precedes the $82 in the example. In the expressions for cents and percent, the words *cents* and *percent* are used. The dollar sign is the only universally approved symbol in technical writing. Note also that modern usage has eliminated the decimals and zeros from even-dollar sums of money. The zeros are added when even-dollar sums and sums including dollars and cents appear in the same list or as items in a series. In the following sentence, both sums are even-dollar amounts; the decimal and zeros are omitted.

**EXAMPLE**

**The cost included $25 for parts and $15 for labor.**

In the next sentence, one number includes dollars and cents; the other number includes even dollars. The decimal and zeros are added to the even-dollar amount.

**EXAMPLE**

**The costs included $29.95 for parts and $15.00 for labor.**

What should be done if one list in a report contains only even dollars, but a later list contains even dollars as well as dollars and cents? The format of the report will be more attractive and the reader will understand the subject matter more easily if decimals and zeros follow all even-dollar sums.

- Numerical approximations are written as words. Approximations of one hundred or more are assumed to be expressed in round numbers of hundreds or thousands. The number in the following sentence is expressed in even hundreds.

**EXAMPLE**

**Nearly two hundred doctors in Boston called for a nuclear-test ban.**

If the statement read "Nearly 175 copies of the report will be needed," 175 is not expressed in even hundreds and is, therefore, written as a figure. In the next sentence, *three thousand* is expressed in even thousands.

**EXAMPLE**

**About three thousand board feet of lumber will be used in the building.**

One exception to this rule is important in technical writing. All technical and scientific measurements, both specific and approximate, are written as figures unless they begin a sentence.

**EXAMPLE**

**This resistor has approximately 1500 ohms of resistance.**

Very large round numbers of millions or billions are usually expressed in a combination of figures and words.

**EXAMPLE**

**The Milky Way contains about 100 billion stars.**

Large sums of money that end in six or more zeros are also expressed in a combination of figures and words.

<table>
<tr><td>

**EXAMPLES**

</td><td>

—$10 million or 10 million dollars

—$1 billion or 1 billion dollars

</td></tr>
</table>

- During the past twenty years, omitting the comma from four-digit figures has become increasingly popular and will probably become a rule. The omission of the comma is justified because numbers from one thousand to nine thousand (except the even-thousand numbers) are usually read as hundreds. Many four-digit numbers such as those used in Social Security numbers, telephone numbers, and dates have always been written without commas. Therefore, the comma is omitted from all four-digit numbers unless the writer has a specific reason for using it. All numbers, however, that have five or more digits have commas after each set of three digits from the right-hand side of the figure.

<table>
<tr><td>

**EXAMPLES**

</td><td>

5000   2795   24,785   193,000   1,789,654

</td></tr>
</table>

- When a number less than a whole number is used, a zero is placed before the decimal. This zero eliminates the possibility of misinterpretation.

<table>
<tr><td>

**EXAMPLES**

</td><td>

0.753   0.005   0.925   0.50

</td></tr>
</table>

If the numbers appear in a column, the decimals must be directly below one another.

<table>
<tr><td>

**EXAMPLE**

</td><td>

0.753
0.0005
0.925
0.50

</td></tr>
</table>

- Figures are always used for numbers in dates; the day usually follows the month.

**EXAMPLE**    July 5, 1981

Notice that the day of the month is separated from the year by a comma. If only the month and year are given, as in July 1981, no comma is needed. Notice also that -st, -nd, -rd, or -th does not follow the day.

- In addresses, the house number is always written as a figure. If the street or avenue has a number below 100 as a name, the street name is written as a word. If the street name is 100 or more, figures are used.

**EXAMPLES**
—3293 South Twelfth East
—195 Fifth Avenue
—296 East 2400 South

Some localities prefer that their streets and avenues be written as figures. A person moving to or doing business in these localities should continue to write out the street names as shown in the example above until it is determined whether figures are preferred.

- In technical writing, specific time is expressed in figures. These figures are followed by a.m., p.m.; or A.M., P.M. Lowercase letters are used more frequently than capital letters. No space is used between the period after *a* or *p* and the *m*.

**EXAMPLES**    7 a.m.    7 p.m.    or    7 A.M.    7 P.M.

As in the rule concerning sums of money, no zeros are used for even hours unless they appear in a list of series of time that also includes minutes.

**EXAMPLES**
—The different shifts begin at 8 a.m., 2 p.m., and 9 p.m. (No minutes are expressed.)
—The different shifts begin at 8:00 a.m., 2:30 p.m., and 9:00 p.m. (Minutes are shown in 2:30 p.m.)

- Arabic numerals are usually used for numbering graphs, charts, or figures.

Arabic or Roman numerals may be used for tables. Roman numerals are used for sections. All the preceding terms are capitalized when they are followed by a specific number. The word *page* is not capitalized. Some specific examples follow.

**EXAMPLES**
—On page 1, Chapter 5, see Figure 1.
—Table I (or 1) is on page 7, Section II.

- In technical writing, capitalize lesser numbered or lettered references such as *step, experiment, test, equation, run, part, area, motor,* or *brand.*

**EXAMPLES**
—After the rotameter was reset, Steps 5 through 7 are repreated.
—By applying Equations 11 and 13 to the raw data from Run 1, we obtain the results shown in Table 3.
—In our laboratory tests, Brand X had higher food value than Brand Y.
—The rental cost for Area C, located on the fourth floor, is $1000 a month.

In recent years, some companies and publications have begun to use lowercase letters for these references. Capitalization, however, is still the preferred form for technical communications unless company policy specifies lowercase letters.

- Names of centuries are usually spelled out. They are not capitalized.

**EXAMPLE**
During the twentieth century, computers have become increasingly important.

- Ages are written as numbers.

**EXAMPLES**
—The 10-year-old machine still functions efficiently.
—These parts are 8 years old.

- The practice of stating a number both as a figure and as a word is used only in legal documents; it is never used in technical writing. If the writer has chosen the most meaningful form, repetition is unnecessary.

---

**EXAMPLE**

**The motor uses 500 watts of power.**

---

The sentence should not be written as "The motor uses five hundred (500) watts of power."

# EXPRESSING FRACTIONS

Because technicians are usually involved in work that requires extensive use of fractions, four guidelines pertaining to fractions are discussed in this section. These guidelines, however, in many ways duplicate the guidelines for the correct use of all numbers.

- All fractions that express general ideas or approximations are written as words.

---

**EXAMPLES**

—**Nuclear energy plants supply approximately one half of the electricity used in Virginia.**

—**The I–215 Belt Route is projected to extend around three fourths of Salt Lake City.**

—**The acid was added until the tube was one-third full.**

---

In the first two sentences, no hyphen is used between the two parts of the fraction; in the third sentence, a hyphen is used. The correct spelling of fractions is explained in "Temporary Hyphens" in Unit 24.

- Fractions used as specific mathematical terms are, like other specific numbers, written in figures.

---

**EXAMPLE**

**The wire is 1/2 inch (or 1/2 of an inch) long.**

---

Either *1/2 inch* or *1/2 of an inch* is correct; *1/2 in.* or *1/2"* is incorrect.

- Fractions are often used as part of compound adjectives. When they are, they must be connected by a hyphen to the adjective following them.

<table>
<tr><td>

**EXAMPLES**

</td><td>

—Use a 1/2-inch drill bit for making the holes.
—The dial has a 1/4-inch bore hub.

</td></tr>
</table>

- In mixed numbers (a combination of whole numbers and fractions), a hyphen or a space separates the whole number from the fraction. The application of this rule avoids misreading, especially when fractions written on a typewriter are made from whole numbers separated by slant lines, as in 3/5 or 5/8.

<table>
<tr><td>

**EXAMPLE**

</td><td>

The test strip was 2-3/8 inches long and 1-1/16 inches wide.

</td></tr>
</table>

If a mixed number is used as part of a compound adjective, the space rather than the hyphen is used; a hyphen is then placed between the mixed number and the adjective that follows.

<table>
<tr><td>

**EXAMPLES**

</td><td>

—The 1 1/8-inch pipe should be used.
—The 4 1/2-inch measurement was inaccurate.

</td></tr>
</table>

Consistency is important in this rule. Either hyphens or spaces should be used throughout the report.

## SUMMARY

1. Rules for using numbers in technical text vary widely, not only from textbook to textbook but also from one kind of an organization, such as a business, to another, such as a field of science that uses exact figures extensively. Technical writers, therefore, must be willing to change number-writing techniques if necessary to conform with any company rules.

2. One guideline concerning numbers never changes: Figures cannot be used at the beginning of a sentence. Unless the number can be expressed in three words or less, the sentence must be revised.

3. Numbers written as figures are more emphatic than those written as words. A writer must use judgment to determine when emphasis is important.

4. One inclusive guideline for numbers specifies that general numbers below 100 are written as words. (Some books say "below 10.") Numbers used to express specific data of any kind are written as figures.

5. An effective technical writer establishes and maintains a consistent method of presenting numbers.

UNIT **26**

# SUGGESTED ACTIVITIES

A. Rewrite any of the following sentences in which numbers are used incorrectly. In a statement following the rewritten sentence, justify your decision. Write the guidelines in complete sentences that are carefully worded and that are grammatically correct. If a sentence as stated is correct, write C; then, no guideline is required.

1. Figure 9–3 reveals that a maximum flux is cut at ninety degrees and at 270 degrees.

2. Twenty-seven dollars was spent for one twelve-volt battery.

3. The employees arrived at 7 a.m., but no one began work until 1:00 this afternoon.

4. The motor uses about 500 watts of electric energy.

5. The company should manufacture about two hundred and fifty microwave ovens this year.

6. The supervisor reported this morning that one-fourth of the drill bits had been sold.

7. We need drill bits for 1/2 inch holes, 3/4 inch holes, and 1 1/2 inch holes.

8. The silicon atom, because it has three shells and 14 electrons, is more complex than a lithium atom, which has 2 shells and 3 electrons.

9. The statement included the following costs: $25 for wire, $15.69 for resistors, $17.80 for tubes, and three dollars for capacitors.

10. The cabinet was three feet high, 10 feet long, and 18 inches deep.

11. They bought 25 two-watt resistors.

12. The 25,000-ohm resistor gives twenty-five times as much opposition to the flow of current as a 1,000-ohm resistor does.

13. A .01 capacitor was needed.

14. The capacitors are listed in the catalog as
.1
.01
.001
.0001

15. James Watt, for whom the watt is named, was born on January 19th, 1736.

**B.** Many guidelines have been given in this unit. Write an original, grammatically correct sentence illustrating each of these guidelines. Use information related to classroom subjects.

**C.** Because many rules for the use of numbers do not have universal acceptance, methods of expressing numbers vary from one book or article to another. Observe how numbers are expressed in different texts, reports, and technical articles. Then, in classroom discussion, consider the following questions:

1. Are the numbers meaningful?
2. Do the numbers distract the reader from other important information?
3. Do the numbers express approximations when specific numbers could be used?
4. Do the numbers follow the rule of consistency?

# Capitalization, Symbols, Abbreviations

## OBJECTIVES

After reading this unit, you should be able to do the following tasks:

- Use the guidelines for capitalization in a written report.
- Use the guidelines for symbols in a written report.
- Use the guidelines for abbreviations in a written report.

## OVERVIEW

Guidelines that answer all questions concerning the correct use of capital letters, symbols, and abbreviations are so numerous that including a complete list in this unit is impractical. The guidelines explained here are those that apply specifically to technical writing and other business communications. Dictionaries, handbooks of grammar usage, and style manuals are excellent references for an extensive study of rules not listed in this text. See the Suggested Supplementary Readings for each section.

# CAPITALIZATION

Capital letters are used in technical writing in three general situations:

1. To begin sentences
2. To distinguish a specific person, place, or thing from a general group of persons, places, or things
3. To distinguish important words in titles and headings from less important words

Many specific guidelines are discussed in this unit, but they are simple rules that have few exceptions.

Consistent application of these guidelines is fundamental to the use of standard English. The current practice adopted by some companies, persons, and publishers to write names and titles entirely in lowercase letters is discouraged in all kinds of formal communication.

## Beginning Words

- Capitalize the first word in every sentence.

**EXAMPLES**

—**Every tool has been inventoried.**
—**Seventy-five percent of the employees received 5-percent raises.**

- Capitalize a quotation that includes a complete sentence or the beginning words of a sentence.

**EXAMPLE**

**The anonymous quotation "Write, not that the reader may understand if he will, but that he must understand whether he will or not" is significant to technical writers.**

If the quotation starts in the middle of the original sentence, the first word of the quotation is not capitalized.

**EXAMPLE**

**The speaker said that the tone used in writing "can build harmonious relations between companies or damage relations already established."**

# Specific Names

- Capitalize specific names of people. (Nicknames are also capitalized, but nicknames are never used in technical writing.)

| | |
|---|---|
| James Watt | Thomas John Jackson |
| William Driggs | Martha Peterson |

- Capitalize specific titles that precede or follow a person's name.

| | |
|---|---|
| Professor Henry Jones | Michelle Smith, Chief Engineer |
| Dr. Joyce Wills | Warren Feldman, M.D. |

- Do not capitalize the names of professions when they are not part of a title.

—That person is a teacher (doctor, lawyer, engineer).
—The research engineer approved the drawing.

- Capitalize titles used as substitutes for names.

| | |
|---|---|
| Mrs. Chairperson | Mr. President |

When *President* is capitalized but does not function as a title or is not preceded by Mr., it refers only to the President of the United States. The article *the* is not capitalized when it precedes a proper noun unless it is clearly a part of the title, name, or organization.

- Capitalize abbreviations of academic degrees.

| | |
|---|---|
| Carlos Ortiz, Ph.D. | Martha Smith, M.A. |

Note that the two parts of the abbreviations are not separated from each other by a space.

- Capitalize names of cities, states, countries, and geographic areas.

**EXAMPLES**

| | | |
|---|---|---|
| Brussels | Chicago | Iowa |
| North Carolina | Europe | Mexico |
| the Northwest | the South | the Mohave Desert |
| the Panama Canal Zone | the Ohio River | the Great Salt Lake |
| the Catskill Mountains | | |

- Do not capitalize a plural generic term that follows more than one specific name.

**EXAMPLES**

the Mohave and Sahara deserts        the Hudson and Missouri rivers

- Capitalize names of historic events such as wars, special days, and eras.

**EXAMPLES**

| | |
|---|---|
| World War I | Independence Day |
| the Renaissance | the Space Age |

- Capitalize names of holidays.

**EXAMPLES**

| | | |
|---|---|---|
| New Year's Eve | Memorial Day | Veterans Day |

*Note:* When *day* is not part of the name, *day* is not capitalized, as in Christmas day.

- Capitalize the names of official government bodies and documents.

**EXAMPLES**

| | |
|---|---|
| United States Senate | the County Commission |
| the State Legislature | the Declaration of Independence |

- Capitalize trade names but not the common nouns that may follow the trade names.

**EXAMPLES**

| Heathkit receiver | Viking 500 transmitter |
| Pacemaker drill | |

- Capitalize the names of races and nationalities.

**EXAMPLES**

| Indian | Oriental |
| American | Russian |

- Capitalize the names of companies, clubs, and organizations.

**EXAMPLES**

| Tektronix, Inc. | the Faculty Club |
| American Business Communications Association | |

- Capitalize the names of schools, including the schools that are part of a university.

**EXAMPLES**

—Westbrook High School
—Hartford Elementary School
—Danville Technical College
—University of Virginia, School of Business

- Capitalize the names of days and months.

**EXAMPLES**

| Monday | Friday |
| October | December |

- Do not capitalize the names of the seasons.

**EXAMPLES**

| summer | winter |
| autumn | spring |

- Capitalize the names of specific classroom subjects.

**EXAMPLES**

Mathematics 102                                Electronics 106

- Do not capitalize general references to classroom subjects unless the subject is the name of a language.

**EXAMPLES**

chemistry                                physics

electromechanics                                English

- Do not capitalize general physical and chemical terms; however, if a person's name is part of these terms, the person's name is capitalized.

**EXAMPLES**

—Boyle's law

—Einstein's general theory of relativity

- Do not capitalize chemical elements and compounds, but capitalize and write without periods the symbols representing elements and compounds.

**EXAMPLES**

water ($H_2O$)                                sodium chloride (NaCl)

cobalt (CO)

- Capitalize *Fahrenheit* and *Celsius*, named for the men who developed these scales for measuring temperature. Do not capitalize *centigrade*. The abbreviations for Fahrenheit (F) and for Celsius or centigrade (C) are capitalized.

- Capitalize the names of astronomical bodies except *earth*, *sun*, and *moon*. *Earth*, *sun*, and *moon* are usually capitalized, however, when used in association with other astronomical bodies.

**EXAMPLES**

—The earth's two great forces are gravitation and magnetism.

—Venus, Mercury, Mars, and Earth are planets; the Sun is a star.

- Do not capitalize most nouns derived from the names of people, nations, or languages.

| | | |
|---|---|---|
| **EXAMPLES** | watt | braille |
| | utopia | hertz |

Some adjectives may be written in either capital or lowercase letters.

| | |
|---|---|
| **EXAMPLES** | —Arabic numerals or arabic numerals |
| | —Roman numerals or roman numerals |
| | —India ink or india ink |

Some adjectives are always written in lowercase letters.

| | | |
|---|---|---|
| **EXAMPLES** | herculean | pasteurized |

Check a standard dictionary to determine whether other words similar to those given in the examples are written in capital or lowercase letters.

- Capitalize the words *figure, table, graph, chapter, section,* and similar references when they are followed by a specific number. Do not capitalize *page.*

## Important Words

- Capitalize the first, last, and all important words in a title of a report, article, or book. The important words include all words except prepositions, conjunctions, and the articles *a, an,* and *the.*

| | |
|---|---|
| **EXAMPLES** | —A Physical Description of the Microwave Oven |
| | —How to Prepare a Contour Map |

This guideline applies only to titles used within context. A report title is usually written entirely in capital letters.

- Capitalize the first word of items in an outline.

**EXAMPLE**

I. Introduction
   A. Background
   B. Purpose
   C. Definitions
   D. Major topics
      1. The body
      2. The handle

- Capitalize the first word of items written in a list. Unless each item of the list is a complete sentence, no punctuation is used.

**EXAMPLE**

The following materials are needed for the experiment:
1. One multimeter
2. One audio-signal generator
3. One resistance decade box
4. Two sheets of linear graph paper

If these items are numbered but written horizontally, they are not capitalized but are separated from one another by commas.

**EXAMPLE**

The following materials are needed for the experiment: (1) one multimeter, (2) one audiosignal generator, (3) one resistance decade box, (4) two sheets of linear graph paper.

# SYMBOLS

Symbols are usually not recommended for use in formal writing. The following guidelines explain when symbols are and are not acceptable in reports and letters.

- Only the symbol for dollars ($) has universal acceptance. It is used every time a specific amount of money, except even millions and billions, is written in formal communications.

**EXAMPLES**

—The total cost of the new electric furnace was $10,750.

—The company plans to construct a 2-million dollar (or $2-million) research plant.

- The percent symbol (%) and the degree symbol (°) are used by many writers but are not universally approved. These two symbols are permitted when both the writer and the reader readily understand them.

- All standard symbols for weights, amounts, quantities, and other mathematical and technical measurements may be used in place of abbreviations or words on graphic aids if space is limited.

- Mathematical terms indicating multiplication, division, subtraction, and addition are expressed as words.

**EXAMPLES**

| | |
|---|---|
| Addition | 25 and (or plus) 19 |
| Multiplication | 4 by (or times) 10 |
| Division | 75 divided by 15 |
| Subtraction | 42 minus 8 |

- The *and* symbol (&) is never used in general context. It is used only as part of a company's name when the originating company shows a preference for the symbol.

**EXAMPLES**

—H & H Engineering

—B and B Steel Fabricators

- The number symbol (#) is never used in context. The capitalized abbreviation *No.* precedes a specific number.

# ABBREVIATIONS

Abbreviations are used in technical reports only when they are more familiar to the reader than the word would be or when the repetition of a long name or phrase sounds repetitious or cumbersome. The few abbreviations that are permitted must be spelled, capitalized, and punctuated correctly. (Correct forms for standard abbreviations are listed in the back section of some dictionaries. More frequently, they are listed alphabetically in the general text.)

The following guidelines help eliminate excessive or incorrect use of abbreviations in technical communications.

- Measurement terms such as inches, feet, yards, meters, liters, gallons, and acres are spelled out.

**EXAMPLES**
—The specifications state that the connecting hose must be 3 feet 2 inches long.
—The space requires 3 cubic yards of concrete.

- Technical terms including complex technical measurements and quantities such as *kilograms, millimeters,* and *megacycles* are written as words unless any one term is used frequently throughout the report. For example, if *millimeter* is a word used many times, it is written out once; the abbreviation is shown in parentheses. The abbreviation is then used throughout the rest of the report. An *s* is never added to the abbreviated plural. The abbreviation is usually written in small letters.

**EXAMPLE**
A meter contains 1000 millimeters (mm).

Simple forms of technical measurements such as *volts, watts, cycles,* and *meters* are always written as complete words.

A good writer learns to use judgment in applying this guideline. If a reader will recognize the abbreviation more quickly than the full word, an abbreviation may be used. However, many abbreviated words following in rapid succession confuse a reader and make the report seem sketchy.

- Abbreviations (or symbols) may be used in graphic aids if space does not permit the word to be spelled out neatly and clearly.

- Some abbreviations are always substituted for the word. These include *Dr., Mr., Ms.,* and *Mrs.* when they are followed by a name, and *No.* when it is followed by a specific number.

**EXAMPLES**
—Dr. James Allen is the company doctor.
—Lot No. 21 is correct.

When two specific numbers are used, they may be written as "Nos. 35 and 42" or as "No. 35 and No. 42."

- Abbreviations that have become more familiar than the word they represent may be used. The list includes FM, TV; college degrees (M.A., Ph.D.); and national organizations (NAACP, NASA, FBI). Only the writer of a specific report can determine when abbreviations of this kind should be substituted for the word.

■ Foreign abbreviations such as *et al.*, *viz.*, and *e.g.* are no longer used in any technical writing. *Etc.* is used only when it follows a list that, if completed, would be unnecessarily long. Even *etc.* (and its English equivalent *and so forth*) can be avoided if the list is prefaced with *such as*, *including*, or a similar introductory word. However, none of these procedures are used if an incomplete list results in the omission of important information.

**EXAMPLES**

—A technical student studies algebra, geometry, calculus, etc.

—A technical student studies many kinds of mathematics, including algebra, geometry, and calculus.

In these sentences, a complete list of mathematics courses is probably not necessary. The following sentence, however, is useless to the reader. A complete set of instructions is needed.

**EXAMPLE**

When loading a revolver, place an explosive cap on each nipple, draw back the hammer to half cock, place a charge of powder in one chamber, etc.

■ The days of the month and the months of the year are spelled out; the names of streets, cities, states, countries, and all similar locations are also spelled out.

**EXAMPLES**

—Testing began on Friday, July 9, 1986.

—Delmar Publishers Inc.

—Two Computer Drive, West

—Box 15–015

—Albany, New York 12212

■ The words *figure, chapter, section,* and *page* are spelled out in technical reports.

**EXAMPLE**

Figure 1 in Chapter 5, page 40, illustrates the valence of hydrogen.

■ Contractions of two words to form a single word are never used in technical reports. They may, however, be used in letters and informal memos.

**EXAMPLES**

| | | |
|---|---|---|
| can't | doesn't | weren't |
| they're | it's | |

Notice that the apostrophe is used in place of the omitted letter or letters.

# SUMMARY

1. All guidelines for capitalizing words are contained in three general statements: Capitalize the first word of *all* complete sentences; capitalize to identify specific persons, places, and things; and capitalize to identify important words in titles and headings. All other, more specific rules fit into one of these categories.

2. Symbols must be used sparingly in technical writing; use them only when there is no doubt that all readers will understand them. The guidelines related to symbols given in this unit should be observed.

3. Technical writers should use abbreviations cautiously, making sure that abbreviations do not follow one another in rapid succession and that the reader immediately understands them.

4. No general guidelines that specify when to capitalize abbreviations are available. Writers must rely upon a good dictionary for this information. (Dictionaries also indicate whether periods are used in the abbreviation.)

UNIT

# SUGGESTED ACTIVITIES

A. Some of the following sentences contain errors in the use of capital letters, symbols, and abbreviations. Correct the errors and tell what guideline was used in making the correction. If the sentence is correct, write C after it.

 1. The elements Hydrogen and Lithium have a valence of 1.

 2. The Chemical Engineer at Acme Prod. Co. discussed the importance of Chemistry in modern industry.

 3. The following apparatus is needed for the experiment:
    1. a glass tube 4″ long
    2. a rubber stopper that fits the tube
    3. a water reservoir
    4. etc.

4. The primary difference between this circuit and the one shown in figure 5 is the 1500-Ω resistor and the filter bypass capacitor.

5. Francis Bacon said, "The ill and unfitting choice of words wonderfully obstructs the understanding."

6. On August 22th, 1980, the Company will move its offices to 36 No. State str.

7. Dearborn, Mich. is the home of Ford motor co.

8. The tools needed to remove the wheel are (1) a jack, (2) a lug wrench, and (3) two wheel blocks.

9. The circuit current is about 0.75 amps; the voltage is about 6 V's, the circuit power is 0.75 $\times$ 6, or about 4.5 w.

10. The Technician inserted 6-volt, 250-ma. lamps into the lamp socket.

11. Christine Adams, chief engineer, approved the Blueprints for the new Highway that will cross the Green and Colorado Rivers.

12. In figure #7, Parallel Constructions are illustrated.

**B.** In a standard dictionary, find the correct abbreviations for each of the following terms.

1. ampere
2. Bachelor of Arts
3. alternating current
4. atomic weight
5. basal metabolism
6. candlepower
7. cycles per second
8. electroencephalogram
9. electromotive force
10. horsepower
11. kilocycles
12. kilowatt-hour
13. microfarads
14. National Academy of Sciences
15. power

**C.** The following paragraph contains numerous abbreviations and a symbol. Assuming that you are writing this text for an uninformed reader, determine whether the paragraph is acceptable as written. If not, rewrite the paragraph and be prepared to justify your revisions. If necessary, use a standard dictionary to determine the words for which abbreviations are used.

The amount of resistance and current must be known before the size of the carbon resistor to be used in the circuit is determined. For example, as shown in Fig. 10, a resistor capable of limiting current through the 900-$\Omega$ resistance to 0.1 amps when the power source is 100 V is inserted in series with $R_1$. Because the total required R is 1000 $\Omega$, the 10-$\Omega$ resistor $R_2$ is connected in series with $R_1$. The current through both $R_1$ and $R_2$ is then 0.1 amp. The $I^2R$ power dissipated in $R_2$ is 1 W, but a 2-W resistor is recommended so carbon resistors do not become overheated during normal use.

**D.** As an alternative to C, students in other fields are encouraged to bring similarly abbreviated text from their area of interest to class for evaluation.

# Correct Use of Words

After reading this unit, you should be able to do the following tasks:

- Use the correct verb and noun.
- Use the correct adjective and adverb.
- Use the correct preposition and conjunction.

## OVERVIEW

No one word in the English language means exactly the same thing as any other word. Even a synonym rarely, if ever, expresses exactly the same idea as the word it replaces. Ideally, English has a right word to express every meaning. Few people, however, are knowledgeable enough to always choose the right word. Nevertheless, every technical writer needs to develop word consciousness because correctly chosen words — when combined with correct grammar, spelling, and sentence structure — help create meaningful, forceful, and convincing reports.

In this unit, many words that are used incorrectly or that have become inadequate synonyms for other words are presented and defined. Most of the words are classified under six parts of speech — verbs, nouns, adjectives, adverbs, prepositions, and conjunctions. A few are miscellaneous.

The necessarily limited information in this unit is given to help students become aware that accuracy in word choice is essential to accurately recorded information. It is not a substitute for a dictionary and a thesaurus, the two essential tools for every technical writer. These tools also help writers select, from a list of synonyms, the word that is most precise and most likely to keep a reader interested.

# VERBS

## Advise, Tell, Inform

*Advise* is not a substitute for *inform*. Advise means, primarily, to caution or to warn. Doctors and lawyers *advise* their patients. One employee, even an executive, *informs* other employees. *Tell* is a mild word that simply means to utter or to relate. *Inform*, of course, means to give information and is the one of the three that should be used extensively in business communications, including technical reports.

**EXAMPLE**

The supervisor *informed* the employees that the equipment must be reordered.

## Affect, Effect

Perhaps no other two words in the English language are more easily confused than *affect* and *effect*. Except in psychology, *affect* is always a verb. As a verb, it means to alter, influence, or change. *Effect* may be a verb meaning to result in, accomplish, or bring about. *Effect* may also be a noun; then, it means the result or the outcome.

**EXAMPLES**

—The weather often *affects* (influences or alters) communication systems.
—The installation of the computer *effected* (resulted in or brought about) a major change in the company's efficiency.
—The *effects* (results or outcomes) of electricity are exciting.

## Aggravate, Irritate

*Aggravate* means to make worse; *irritate* means only to annoy. One is not a substitute for the other.

**EXAMPLES**

—The delay in repairing the machine *aggravated* the problem (made it worse).
—The noise of the machine *irritated* (annoyed) the employees.

## Aim, Intend

*Aim* means to point a weapon or to point toward some objective; *intend* means to plan to do something. The two words are not synonymous.

—The company is *aiming* toward improved communication among employees.
—Jack *intends* to report the error.

## Allege

Although *allege* can mean to assert positively or affirm, it also means to assert without proof. It is a legal term, not a term to be used in technical writing.

## Allow, Think

*Allow* means to permit; it is not a synonym for the verb *think*.

—The foreman *allowed* Bill to dissemble the motor.
—Bill *thinks* that the problem is in the detector circuit.

## Appear, Seem

In its primary sense, *appear* means to come into view. Although in another sense, *appear* approximates the meaning of *seem*, which means to give evidence of being, *seem* is preferred in such sentences as "The machine *seems* to be overloaded" or "The voltage *seems* to be low."

## Appreciate, Understand

*Appreciate* means to value, treasure, or cherish. In technical writing, it is not an acceptable substitute for *understand* when *understand* means only to grasp the meaning of.

—The trainee *appreciates* (values) the opportunity to check the equipment.
—The machinist *understands* the problem and will work on it today.

## Conclude, Decide

Although *conclude* and *decide* are sometimes much alike in meaning, *conclude* means to make a final judgment based upon evaluation of information. *Decide* may simply mean to make a choice. Of the two words, *conclude* has the greater impact and almost invariably is used only to indicate that all facts have been considered before a decision is made.

EXAMPLES —After Gerry checked the tubes, he *concluded* that none were defective.
—The technician *decided* to install a new motor.

## Forward, Send

Although *forward* means to transmit or to send onward, it has been overused in written communications. More specific words such as *mail*, *send*, or *ship* are now preferred.

**EXAMPLE** The company *mailed* (not *forwarded*) the merchandise on June 6.

## Infer, Imply

The reader *infers;* the writer *implies.*

**EXAMPLES** —I *infer* from your report that the machine is broken.
—The technician *implied* that a new picture tube should be installed.

## Lie and Lay

The verb *lie* means to recline or rest in a horizontal position. The forms are *lie, (is) lying, lay, (have) lain.*

The verb *lay* means to put or to place something somewhere. The forms are *lay, (is) laying, laid, (have) laid.*

The verb *lie* does not take an object. Thus, one never lies something down. The verb *lay*, however, may take an object or may be used in the passive voice.

**EXAMPLES** —The concrete *lies* in a form until the concrete hardens.
—Construction workers *lay* insulation between the framing members.
—For a foundation wall, concrete blocks *are laid* below the ground. (Passive voice)

## Obtain, Secure

*Obtain* means to gain possession of; *secure* means to fasten or make safe. *Secure* is not a synonym for *obtain.*

**EXAMPLES**
—*Obtain* the blueprints from the office.
—*Secure* the equipment from fire.

## Rise and Raise

The verb *rise* means to move upward. Its forms are *rise, (is) rising, rose, (have) risen.*

The verb *raise* means to cause or to help something move upward. The forms are *raise, (is) raising, raised, (have) raised.*

The verb *rise* is used when the subject of the verb is itself moving upward. This verb does not take an object. The verb *raise* does take an object.

**EXAMPLES**
—The water level will *rise* before midnight.
—Workers will *raise* the beam several inches tomorrow. (The object is *beam*.)

## Sit and Set

*Sit* means to assume an upright resting position. The forms are *sit, (is) sitting, sat, (have) sat.*

*Set* means to put or to place something somewhere. The forms are *set, (is) setting, set, (have) set.*

The verb *sit* seldom takes an object; the verb *set* can take an object.

**EXAMPLES**
—Have patients *sit* down before you take their temperature.
—He *set* the chart on the table while he worked. (The object is *chart*.)

## Suspect, Suspicion

*Suspicion* is a noun; it cannot be used as a substitute for the verb *suspect.*

**EXAMPLE**
Bill *suspects* (not *suspicions*) that the battery is weak.

## Wait For, Wait On

*Wait on* is a term used only for service given in a restaurant. A person *waits for* people, *waits for* parts, or *waits for* the negatives to dry.

# NOUNS

## Ability, Capacity

*Ability* means having the power to perform. *Capacity* means having the power to receive, hold, store, or absorb.

EXAMPLES
—The tank has a 200-gallon *capacity.* (It can hold 200 gallons.)
—The machine has the *ability* to print five hundred copies a minute. (It has the power to produce or turn out five hundred copies.)

## Data, Datum

*Datum* is the singular form of data but is rarely used. *Data*, the plural form, theoretically requires a plural verb when used as a subject. Dictionaries, however, permit the use of *data* as either a singular or plural form. All the following examples are written correctly.

EXAMPLES
—These *data* are correct; this *data* is correct.
—These are the correct *data;* this is the correct *data.*

## Height, Heighth

*Height* is the only correct form; *heighth* is a misspelled form of *height.*

## Principle, Principal

*Principle* is always a noun. It refers to a set of rules or standards. *Principal* may be a noun or an adjective. As a noun, it refers to a person who is in a leading or an authoritative position. It also refers to money placed at interest. As an adjective, it means leading, important, or influential.

EXAMPLES
—Eight basic *principles* (rules) for technical writing were introduced.
—The *principal* speaker explained *principal* and interest.

# ADJECTIVES

## All Right, Alright

*Alright* at one time was an accepted form. Today, it is only a misspelling for *all right*. Therefore, *all right* is the only correct form; it is always spelled as two words.

## Anxious, Eager

*Anxious* suggests worry or doubt. It should never be used for *eager* when the writer means enthusiastic anticipation.

EXAMPLES
—The chemist is *anxious* (worried) about the results of the experiment.
—They are *eager* to begin the inspection.

## Continual, Continuous

*Continual* means that something goes on over a long period of time but that the activity may be interrupted. *Continuous* means that no interruption takes place.

EXAMPLES
—The company experienced *continual* problems with the computer.
—The *continuous* noise of the machine was distracting.

## Different From, Different Than

*From* is a preposition; *than* is a conjunction. Because *different* is followed by a prepositional phrase in most sentences, *different from* is usually correct. Occasionally, *different* is followed by a clause; then, *different than* is used.

EXAMPLES
—The atoms in one element are *different from* those in another element. (*From those* is a prepositional phrase.)
—The results of the test are *different than* the technician expected. (*Than the technician expected* is a dependent clause.)

## Each Other, One Another

*Each other* is correct when two people or things are involved. *One another* is correct when more than two people or things are involved. If the sentence does not make specific reference to two and only two, *one another* is used.

—X-ray lasers and chemical lasers differ from *each other* in several ways.

—The various kinds of lasers differ from *one another* in several ways.

## Enthused, Enthusiastic

*Enthused* is a verb; even as a verb, however, it is usually avoided in formal writing. Because *enthusiastic* is an adjective, *enthused* can never be substituted for *enthusiastic.*

—The technologist was *interested in* (not *enthused about)* the results of the test.

—The *enthusiastic* employees waited for the arrival of the new computer.

## Fewer, Less

*Fewer* modifies plural nouns; *less* modifies a quantity or a singular noun.

—This machine prints *fewer* copies than the old one did.

—A lithium atom has *fewer* electrons than a silicon atom.

—The new motor produces *less* power than the old one.

*Fewer* and *less* are comparative adjectives. When they are used, a complete comparison must be shown.

## Kind Of, Somewhat, Slightly

*Kind of, somewhat,* and *slightly* are vague adjectives that convey very little meaning to the reader and should usually be avoided in technical writing.

## Lot Of, Lots Of

*Lot of* or *lots of* are poor substitutes for *many.*

*Many* (not *lots of)* electronic devices contain triode vacuum tubes.

Even the adjective *many* should be avoided if a specific number can be used.

**EXAMPLE**

**Seventy-five percent of the tubes have been inspected.**

## Regardless, Irregardless

*Irregardless* is a substandard word that is not acceptable; *regardless* is the correct form.

## Unique, Correct, Perfect, Round

Words such as *unique, correct, perfect,* and *round* cannot be compared. For example, nothing can be more round than round or more unique than unique. Check carefully the meanings of these and similar words to determine whether *more, less, most,* or *very* may logically be combined with them. It is correct to say that something is more nearly unique, round, or perfect than something else when neither item being compared has attained the ultimate rank of uniqueness, roundness, or perfection. It is also correct to say that something is almost unique, almost round, or almost perfect.

**EXAMPLES**

—That drawing is *more nearly perfect* than this one.
—The answer is *almost correct.*

# ADVERBS

## Around, Approximately, Almost

*Around* is not a satisfactory substitute for *about, approximately,* or *almost.* All these words should be used cautiously in technical writing because they express vague ideas, not specific ones.

## Basically, Essentially

*Basically* and *essentially* are frequently misused. They are powerful adverbs that imply something is indispensable to the whole part. For example, the sentence "All electron tubes are basically (or essentially) alike" means that all electron-tube design begins with a single principle—that electrons flow from negative to positive. Because *basically* and *essentially* refer to underlying principles or elements, they cannot correctly substitute for *principally, chiefly, in effect, most of,* or *almost.*

## Farther, Further

*Farther* refers to physical distance—distance that can be measured. *Further* refers to time or degree.

**EXAMPLES**
—This measuring tape extends *farther* than the plastic one does.
—The results of the test require *further* study.

A few recently published English textbooks permit *farther* and *further* to be used interchangeably. Until this practice has become more widely accepted, a distinction should be made between *farther* and *further* in formal writing.

## First, Second, Last

*First*, *second*, and *last* (or *third*) are much preferred to *firstly*, *secondly*, and *lastly* (or *thirdly*); but the numbers *1*, *2*, and *3* are usually preferred to either of the two sets of words.

**EXAMPLE**
To regap a spark plug, perform the following procedures: *(1)* Remove the spark plug from the engine, *(2)* gap the spark plug, and *(3)* replace the spark plug in the engine.

## Liable, Likely

*Liable* indicates responsibility. It also indicates probability tinged with some unpleasantness or danger. *Likely* implies possibility. In precise writing, these two words should not substitute for each other. See the following correctly written sentences.

**EXAMPLES**
—You are *liable* for the equipment in the shop.
—The supervisor is *likely* to ask for your report.

## Respectively, Respectfully

Though widely removed from each other in meaning, *respectively* and *respectfully* are often misused. *Respectively* means in the order given; *respectfully* means out of respect or esteem.

| EXAMPLES | —The boards measure 8 feet, 9 feet, and 10 feet, *respectively.*<br>—They *respectfully* obeyed the instructions from the foreman. |
|---|---|

## Sooner, Rather

*Sooner* means before another event takes place. It is never a substitute for *rather*, which means preferably.

| EXAMPLES | —The drill bits were assembled *sooner* than the company expected them to be.<br>—Pat would *rather* repair a radio than a television set. |
|---|---|

## Than, Then

*Than* is a conjunction; it introduces a comparison. *Then* is an adverb; it answers the question *When?* One is not a substitute for the other.

| EXAMPLES | —Larger and quieter gears are being manufactured today *than* were manufactured in the past.<br>—A survey team will stake out the four corners of the lot and connect them with straight lines; *then* the new island will be located and staked out. |
|---|---|

# PREPOSITIONS

## As To, About, Concerning

*As to*, *about*, and *concerning* as prepositions are synonymous. However, *as to* is used only in reference to ideas; *about* or *concerning* are used in reference to both ideas and objects.

| EXAMPLES | —The technician was consulted *as to* (*about, concerning*) a solution for the problem.<br>—A seminar *about* (or *concerning*) cross tabulation is scheduled for next Monday. |
|---|---|

## Between, Among

*Between* is used when only two people or things are involved; *among* is used when more than two people or things are involved.

EXAMPLES
—More distance is needed *between* the input and the output of the machine.
—Many capacitors were found *among* the resistors in the bin.

## Due To, Because Of

*Due* can be an adjective; *due to* can be a preposition synonymous with *because of*. As an adjective, *due*, which is normally followed by the preposition *to*, must either follow a noun that it modifies or function as predicate adjective following a form of the verb *to be*.

As a preposition, *due to* has not received wide acceptance. Therefore, technical writers should avoid the risk of offending some readers and use *because of* in adverbial phrases that tell why.

**EXAMPLES**
—The electrical failure was *due to* the storm.
—*Because of* the electrical failure, the ovens could not be used.

## Inside Of

*Inside of* is never correct. Say "inside the engine" or "inside the tube."

## Like, As

*Like* may be a verb or it may be a preposition, but it is never a conjunction; therefore, it cannot be used as a substitute for the conjunctions *as if* or *as though*.

**EXAMPLES**
—That motor looks exactly *like* this one.
—It looks *as if* the wrong measurement  was  made.

## Off, Off Of

*Off* is correct; *off of* is incorrect.

**EXAMPLE**
The printed circuit boards were taken *off* the assembly line.

## Per

*Per*, a Latin word, is preferably used only in Latin expressions such as *per diem* or *per annum*. *Per* may also be used to designate a specific rate such as "$25 per day" or "30 rotations per minute." However, "$25 a day" and "30 rotations a minute" are preferred. *Per* is not correct in general reference.

---

**EXAMPLE**

**We shipped the tubes *according to* (not *per*) your request.**

---

# CONJUNCTIONS

## As, Because, Since

*As* is not a meaningful substitute for *because* or *since*. Of the three words, *because* most specifically introduces a clause that answers why. As a conjunction, *since* is preferably used to express time relationships.

---

**EXAMPLES**

**—*Because* (not *as* or *since*) a hydrogen atom has one electron in its outer shell, this atom is said to have a valence of 1.**

**—People have feared nuclear war *since* the hydrogen bomb was invented.**

---

In the second example, note that *since* still has two possible meanings — reason and time. Therefore, *since* is rarely precise as a conjunction. It is precise as a preposition.

---

**EXAMPLE**

***Since* the invention of the hydrogen bomb, people have feared nuclear war.**

---

## Neither, Nor

When *neither* is used as part of a correlative conjunction, it must be followed by *nor*, never by *or*.

---

**EXAMPLE**

***Neither* the voltmeter *nor* the ammeter could be located.**

---

## Not, Or

When *not* is used as an adverb, *or* is used as a conjunction.

| EXAMPLE | The technician could *not* find the voltmeter *or* the ammeter. |

## Provided, Providing

As a conjunction, *provided* is preferred to *providing;* however, both should be avoided when *if* can be used.

| EXAMPLE | The printed circuit board can be inspected *if* it has been completed. |

## Reason Is Because

*The reason is because* is always incorrect. *The reason is that* may be used, but even this wordy expression can usually be replaced by the single word *because.*

| EXAMPLE | The power could not be determined *because* (not *for the reason that*) the voltage was unknown. |

## While

*While* means during the time that. It is neither a substitute for the coordinating conjunctions *but* or *and* nor a precise substitute for the subordinate conjunction *although.*

| EXAMPLES | —Potential energy is energy of position, *but* (not *while*) kinetic energy is energy of motion. <br> —*Although* (not *while*) the copy machine was not functioning, only minor repairs were needed. <br> —*While* the copy machine was being repaired, the company rented one. |

## With

The preposition *with* has a variety of meanings and uses. Among these meanings and uses, it can substitute for *that has, that have,* or *that uses* or *use.* However, in technical communications requiring precise meanings, *that has, that have,* or *that uses* or *use* is preferred to *with.*

—A transistor radio is a radio *that uses* (not *with*) solid-state amplifier devices.
—A person *with* a winning smile is always welcome.

In these examples, *with* implies *together with* and *and*, respectively, in addition to the intended meanings *that uses* and *who has* and thus causes ambiguity.

## Yet

*Yet*, when used as a conjunction, attracts attention to itself and away from the meaning expressed in the sentence. Words such as *however, nevertheless,* or *but,* which have more specific meanings, are usually preferred to *yet.*

# MISCELLANEOUS WORDS

## Kindly, Please

Both *kindly* and *please* are used to transmit polite requests. However, *please,* which implies a mutual relationship, is preferred in technical communications to *kindly,* which implies gratitude.

—*Please* (not *kindly*) consider my request for private office space.
—Will you *kindly* (or *please*) call a cab for me?

## Its, It's

*Its* is the possessive form.

The company gives *its* employees many benefits.

*It's,* a contraction for *it is,* is not used in formal reports because contractions are not permitted. *It's* may be used in letters and informal memos.

## Same

In technical communications, *same* used as a pronoun is not generally accepted and should thus be avoided as a substitute for *it* or for any noun. *Same* is accepted as an adjective.

| EXAMPLES | —The technician removed the transistor and checked *it* (not *same*).<br>—The technician had removed and checked the *same* transistor previously. |

## Said

*Said* is a verb; it is not a substitute for *previously mentioned*.

| EXAMPLE | The *previously mentioned* (not *said*) transistor was not defective. |

## Use To, Used To

Neither *use to* nor *used to* is an acceptable substitute for *formerly*.

| EXAMPLE | We *formerly* (not *used to*) put tubes instead of transistors in these sets. |

# SUMMARY

1. Correctly used words are essential to correctly written communications.

2. The words, explanations, and examples in this unit do not solve all word-choice problems. The list given is intended only to help writers become continually aware that words similar in sound or spelling can have vastly different meanings.

3. Writers should refer to a dictionary or thesaurus to verify the meaning of a word they are questioning or to choose the most precise words from one or more synonyms.

---

UNIT **28**

## SUGGESTED ACTIVITIES

A. Change or eliminate any incorrectly used words in the following sentences. Rewrite the sentence if necessary, but do not change the meaning.

1. The problem appears to be in the output circuit.

2. Technicians used to use lots of different kinds of meters in the laboratory.

3. The research department was advised by the foreman that the new building was essentially completed.

4. That reading on the voltmeter seems to be more correct **than the first one.**

5. The computer is liable to make less errors if the input material is correct.

6. Further tests indicate that the principle cause of error was due to a defective meter.

7. The company concluded that it would sooner buy new equipment than repair the old equipment.

8. The instructor aims to explain the effect of resistance upon electron movement.

9. Conductors offer little opposition to the flow of current while insulators offer much opposition to the flow of current.

10. Metals are the best conductors with copper, aluminum, and iron wire being the metals most frequently used.

11. Providing a length of resistance wire, a switch, and a dry cell are connected in series, the effect of current on resistance can be seen.

12. The resistance of copper wire is different than the resistance of silver, iron, and other metals if they all have identical dimensions.

13. As current is always greatest through the path of least opposition, total current is increased when resistors are parallel to each other.

14. The d'Arsonval movement for meters with the 0–1 milliampere range is essentially a basic type of movement.

15. This data will be used due to the fact that its accuracy has been verified.

**B.** The following list contains words, most of them pairs of words similar in meaning or spelling, that are often misused in technical communications. Study the list carefully, and check at least twenty-five words that you may be likely to use incorrectly. Using a dictionary, determine the meanings of the words you choose; then, use each of the chosen words correctly in a sentence.

| | | |
|---|---|---|
| accept | balance | complementary |
| except | rest (remainder) | complimentary |
| allude | calendar | device |
| elude | calender | devise |
| allusion | capital | discover |
| elusion | capitol | invent |
| illusion | center | formally |
| attain | middle | formerly |
| retain | circular | forth |
| | round | fourth |

| | | |
|---|---|---|
| lessen | mutual | stationary |
| lesson | common | stationery |
| lineal | party | their |
| linear | person | there |
| loose | personal | they're |
| lose | personnel | valuable |
| majority | precede | valued |
| most | proceed | |
| moral | role | |
| morale | roll | |

C. Notice the use of words in textbooks, technical and professional magazines, and newscasts. Many are used incorrectly. Bring to class a copy of sentences containing words that seem to be used incorrectly. Talk about the sentences in a classroom discussion.

# Aids to Conciseness

After reading this unit, you should be able to do the following tasks:

- Use words for phrases.
- Use the correct word length.
- Avoid using unnecessary words.
- Avoid using meaningless words.
- Avoid needless repetition.

## OVERVIEW

A concise style of writing is vigorous and interesting because it permits a sentence, paragraph, or report to say much in a few words. At the same time, it gives the report a polished smoothness that accomplishes the purpose logically and systematically.

A reader expects technical reports to be concise but also clear, accurate, and complete. Technical writers can meet these requirements by applying the following procedures, all of which are discussed in this unit: using words for phrases, using correct word length, avoiding unnecessary words, avoiding meaningless words, and avoiding needless repetition.

# WORDS FOR PHRASES

Wordy phrases in formal writing make a report unnecessarily long, make sentences lose forcefulness, and conceal meaning behind a maze of words. Consider the following sentence.

**EXAMPLE**
Due to the fact that the members of management conducted an investigation with reference to the problems that were developing within the ranks of the employees in the electrical shop, a meeting was called.

This sentence contains thirty-four words. When single words replace phrases, the meaning is expressed clearly and confidently in eleven words.

**EXAMPLE**
Management personnel met to discuss the problems developing among the electricians.

The following examples show how other short, specific words can replace phrases that retard the flow of information.

**EXAMPLES**

| Phrase | Change to |
|---|---|
| a large number | many, some, or a specific number |
| as a whole | entire |
| along the lines | like, similar to |
| am (are, is) in a position | can |
| are of the opinion that | believe |
| at an early date | soon (or give a specific date) |
| at all times | always |
| at the present time | now |
| at a time when | when |
| conduct an investigation | investigate |
| due to the fact that | because |
| effect an improvement | improve |
| give assistance to | aid, help, or assist |
| give consideration to | consider |
| have a preference for | prefer |
| have (hold) a discussion | discuss |
| I would appreciate it if | please |

| Phrase | Change to |
|---|---|
| in the event that | if |
| in order to | to |
| in view of the fact that | because, since |
| is applicable | applies |
| make an inspection | inspect |
| make a study of | study |
| prior to | before |
| with reference to | about, concerning, relating |

# CORRECT WORD LENGTH

The second aid to conciseness is using a short, familiar word every time one can be found to express the intended meaning. Because short words are rarely misunderstood, they transmit information easily and quickly. They also prevent the unclear meanings that result when many long words follow one another in a sentence.

In the following examples, long or pretentious words are changed to short, familiar ones.

**EXAMPLES**

| Word | Change to |
|---|---|
| accorded | given |
| aggregate | total |
| ameliorate | improve |
| apprise | tell, inform |
| commence | begin |
| comprise | include |
| conduct | lead |
| deem | think |
| dwell | live |
| feasible | possible |
| forenoon | morning |
| indicate | tell, show |
| initiate | begin, start |
| interrogate | ask, question |
| majority | most |
| necessitate | require |

| Word | Change to |
|------|-----------|
| presently | now (or soon, depending upon intended meanings) |
| purchase | buy |
| reflect | show |
| specify | name |
| sufficient | enough |
| utilize | use |
| whenever, wherever | when, where |

Short words lose their value, however, if they cause the sentence to become wordy and cumbersome. In the following example, one fairly long word is more concise than several shorter ones.

---

**EXAMPLE**

*Wordy:*  Words *that are used merely to "show off" one's vocabulary* should be avoided.

*Better:*  *Pretentious* words should be avoided.

---

Effective writers develop vocabularies that give them several words to choose from. Then, using judgment, they can determine the word length to be used. Remember to choose words for their ability to aid conciseness, not to impress the reader.

# UNNECESSARY WORDS

An unnecessary word in technical writing is any word that contributes nothing toward transmitting information to the reader. Unnecessary words include weasel words, words following the expletive *it*, weak modifiers, and miscellaneous words and phrases that do no more than increase the length of the sentence.

## Weasel Words

According to Webster's dictionaries, the term *weasel word* has reference to the weasel's habit of removing the contents from an egg while leaving the shell intact. Thus, a weasel word is one a writer uses to deliberately create ambiguity. Weasel words are used by writers who lack confidence in their statements, recommendations, or proposals. These words violate the rules of ethical standards for technical writing because they tend to conceal or to distort facts. Some words and phrases often used as weasel words are shown in the next examples.

| **EXAMPLES** | generally speaking | as has been suggested |
|---|---|---|
| | to be sure | customarily |
| | it appears (or it seems) | in keeping with company policy |
| | normally | as you can see |
| | apparently | as might be expected |
| | supposedly | is considered (or thought) to be |
| | possibly | for the most part |

Notice how the use of weasel words weakens and distorts facts in the following sentences.

**EXAMPLES**
—The total circuit power supposedly should be 4.5 watts.
—It appears that most AC waveforms are sine waves.

Notice also how the tone of confidence in one's self and one's information emerges when the weasel words are omitted.

**EXAMPLES**
—The total circuit power should be 4.5 watts.
—Most AC waveforms are sine waves.

## Words Introduced By *It*

*It* is usually a pronoun but is sometimes an expletive, a word used only to fill up space. Words introduced by the expletive *it* weaken the rest of the sentence because they give no specific information and they introduce the sentence, forcing more important words into less emphatic positions. The following expressions and similar ones are seldom needed.

| **EXAMPLES** | It is obvious (apparent) | It has been shown (illustrated) |
|---|---|---|
| | It is known | It should be noted |
| | It might be necessary | It should be remembered |

The words *note* or *remember* concisely communicate the same idea as the phrases above.

## Weak Modifiers

Sometimes, adverbs such as *most, very, really,* and *surely* weaken rather than strengthen the words they modify and should, therefore, be omitted. Notice how the qualifying adverb in the first sentence of each pair of sentences attracts attention to itself and thus weakens the more important word that follows.

**EXAMPLES**

—The machine is working very efficiently.
—The machine is working efficiently.

—That information was really interesting.
—That information was interesting.

—The experiment was really successful.
—The experiment was successful.

## Miscellaneous Unnecessary Words

The first sentence in the following pairs of sentences contains unnecessary words. The second sentence illustrates how these words can be eliminated.

**EXAMPLES**

*Wordy:*   The supervisor came and inspected the shop.
*Better:*   The supervisor inspected the shop.

*Wordy:*   The work was performed in a way that was efficient.
*Better:*   The work was performed efficiently.

*Wordy:*   The valuable employee is the one who takes great care with his or her work.
*Better:*   The valuable employee works carefully.

*Wordy:*   The tools that are most frequently used in the shop are screwdrivers, pliers, and wrenches.
*Better:*   Screwdrivers, pliers, and wrenches are the most frequently used tools in the shop.
*Or*   The tools used most frequently in the shop are screwdrivers, pliers, and wrenches.

The following words and phrases are unnecessary in letters.

| EXAMPLES | | |
|---|---|---|
| | Please be advised | We remain |
| | This is to inform you | We desire to acknowledge |
| | As per your request | The undersigned |
| | I beg to remain | According to our records |
| | I beg to advise | Referring to the matter |

# MEANINGLESS WORDS

Information in technical writing is specific so that the reader and the writer can make identical interpretations. *Seven inches, 100 percent,* and *20 pounds* are specific; the reader is not likely to misunderstand them. General terms such as *excellent, efficient,* and *substantial increase* may be interpreted one way by the writer and another way by the reader. General words may be used for general references in technical reports, but they are meaningless as substitutes for specific information. The following sentences and revisions illustrate the difference in meaning between general and specific words.

**EXAMPLES**

*General:* The temperature must be checked periodically.
*Specific:* The temperature must be checked every 25 minutes.

*General:* The drop in line voltage caused the motor to draw more current.
*Specific:* The drop in line voltage caused the motor to draw 20 percent more current.

*General:* The use of the computer increased this year's profits considerably.
*Specific:* The use of the computer increased profits $25,000 for 1986.

Other general words that become meaningless when used carelessly in technical writing are listed in the following examples.

**EXAMPLES**

| | |
|---|---|
| a number of | fair |
| appreciable | more or less |
| approximate | negligible |
| character | quite a few |
| comparative | reasonable |
| conditions | relative |
| considerable | sufficient |
| definite | suitable |
| evident | undue |
| excessive | |

# REPETITION

Important concepts in technical communications need emphasis; however, dramatic techniques such as underlining and exclamation marks are discouraged. Carefully planned repetition of words and ideas helps to create this emphasis. In contrast, repetition not used for a specific purpose can be meaningless, wordy, and distracting. Therefore, a distinction between effective repetition, which is informative, and needless repetition, which is disruptive, provides a valuable tool for writers.

## Effective Repetition

Five situations in technical writing require skillful use of repetition.

1. Ideas are repeated in reports to introduce the subject, to discuss the subject, and to summarize the information given.
2. When a specific word is used to describe a process, that word is repeated each time the process is referred to in the report. For example, if a writer says that parts have been *soldered* together, *soldered* is repeated if the process is discussed more than once. Substituting general terms like *connected* or *joined* to avoid repetition is distracting.
3. Repetition is used in parallel constructions, as in the following example.

**EXAMPLE**

AC current rises to maximum then falls to zero in one direction; it then rises to maximum then falls to zero in another direction.

The repetition in these two parallel clauses defines AC current clearly and concisely. Similar sentence structure is recommended for expressing measurements, quantities, weights, and the like in technical writing.

**EXAMPLES**

—The bore diameter is 0.2598 of an inch; the precision-shaft diameter is 0.2497 of an inch.
—$R_1$ has 50 ohms of resistance, $R_2$ has 50 ohms of resistance, but $R_3$ has 75 ohms of resistance.

4. Repetition of a noun is necessary when a pronoun does not clearly indicate the word being referred to.

**EXAMPLE**

When a force is used to move an object, it is expended.

The meaning of *it* is not clear; therefore, *force* must be repeated: "When a force is used to move an object, force is expanded."

Here is another example.

---

**EXAMPLE**

**In test No. 2, the line voltage was applied to the motor to permit it to drive the generator.**

---

A reader must guess what *it* refers to — voltage or motor. Repetition of the word *motor* makes the meaning clear: "In test No. 2, the line voltage was applied to the motor to permit the motor to drive the generator."

5. Vague pronouns including *this, that, these, those, it, they, he* and *she* can never be used at the beginning of a paragraph unless they are immediately followed by a noun that clarifies the meaning of the pronoun. This noun is repetition of the noun used in the preceding paragraph.

## Needless Repetition

Needless repetition results from careless choice of words or from inadequate vocabulary. Three kinds of repetition are confusing and wordy.

1. When a form of a word being defined is used to explain the word, the definition has no value. Consider the following sentence.

---

**EXAMPLE**

**A computer is a machine designed to make mathematical computations.**

---

The person who needs a definition of *computer* also needs a definition of computation. The definition has said nothing meaningful.

2. A word or a different form of the word repeated in a sentence often results in needless repetition.

---

**EXAMPLE**

**On each side of the thermometer are numbers that are used to designate degrees of temperature.**

---

The repetition of *are* is unnecessary. One revision is "On each side of the thermometer are numbers that designate degrees of temperature."

Here is another example.

---
**EXAMPLE**
---

**The electric motor used 500 watts of electricity.**

In this sentence, the motor is understood to be an electric one: "The motor used 500 watts of electricity" eliminates needless repetition.

3. Needless repetition occurs when an adjective modifying a noun or an adverb modifying a verb repeats the meaning expressed by the noun or verb. This repetition occurs when a writer does not study the meanings of words carefully. For example, in the expression "wholly integrated," *integrated* means to make into one. To "wholly make into one" is redundant. Similar expressions in which the modifying words or phrases are unnecessary are listed in the following examples.

---
**EXAMPLES**
---

| | |
|---|---|
| *absolutely* essential | *completely* finished |
| *actual* experience | *completely* oriented |
| cancel *out* | few *in number* |
| collaborate *together* | first *of all* |
| *completely* eliminated | join *together* |

In some expressions, duplicate nouns or phrases are repetitious. See the following examples.

---
**EXAMPLES**
---

| Phrase | Change to |
|---|---|
| each and every | each (or every, but not both) |
| in full force and effect | in force |
| cost the sum of | costs |
| in the year of 1965 | in 1965 |
| in the city of Seattle | in Seattle |
| in the state of Maine | in Maine |
| for the period of a year | for a year |
| during the course of | during |
| whether or not | whether |

# SUMMARY

1. Conciseness is a valuable quality in technical writing. It saves readers' time and usually contributes to clarity. (It also lessens the need for extensive punctuation.)

2. A writer striving for conciseness never sacrifices completeness, clarity, or accuracy.

3. Conciseness is achieved through replacing phrases with words; using short words whenever possible; choosing words that, as used, have a single meaning; and using repetition skillfully.

4. Writers who merely adhere to guidelines are not likely to develop communications that are concise, clear, complete, and accurate; judgment and awareness are also essential.

5. Awareness of wordiness in text written by others helps writers avoid wordiness in their writing.

6. A writer's critical review of text he or she has written hours or days earlier helps develop conciseness.

---

UNIT

# SUGGESTED ACTIVITIES

A. Rewrite the following sentences to make them more concise; be sure to retain the original meaning:

1. The invention of the central grid initiated the development of the electronics industry.
2. One of the important uses of the cold-cathode tube is its use as a voltage regulator.
3. It is apparent that there must be an accompanying plate-voltage drop when plate current flows through a diode.
4. A large number of elements are classed as metals.
5. The project engineer considers it feasible to originate a training program for new employees.
6. The two electric motors both operate on the same horsepower.
7. Gold is a dark yellow in color.
8. The mathematician bisected the angle into two equal parts.
9. Originally, the ores were of igneous origin.
10. It should constantly and continuously be borne in mind that a sufficient drop in voltage will cause a motor under load to draw excessive current and become damaged.

11. Injury to the kidneys may be caused by the deposition of insoluble material in the tubules, the interstitium, or the urinary tract.

12. The investigation team recommends that consideration be given to modernization and beautification of the expanded parking lot.

13. The actual facts are that the dispute cannot be resolved without very substantial salary increases.

14. In most practical applications there will probably be some slippage in the belt.

15. Because of inadequate drainage, a sudden inundation of the land is likely to occur.

16. There is a probability that some of the veins may have had the content of their gold increased by enrichment.

B.  In a dictionary or thesaurus find shorter or more familiar words for the following:

| | |
|---|---|
| accede | discontinue |
| accommodate | endeavor |
| activate | expedite |
| affix | facilitate |
| alleviate | function (verb) |
| ascertain | implement |
| category | instigate |
| cogitate | numerous |
| cognizant | subsequent to |
| compensation | terminate |
| conflagration | veracious |
| demonstrate | |

C.  Write ten sentences containing technical information. Wait for two or more hours; then, revise the sentences. Look for ways to make the sentences more concise without omitting information.

D.  Make a notebook of sentences found in your reading that can be written more concisely. Write the original sentence and a possible revision. Look for unnecessary words, phrases that can be changed to single words, meaningless words, and needless repetition.

# 30

# Aids to Clarity

## OBJECTIVES

After reading this unit, you should be able to do the following tasks:

- Use meaningful transitions in writing.
- Write paragraphs correctly.
- Use lead-in sentences in paragraphs.
- Use concrete words and action verbs in sentences.
- Omit negative expressions and substandard expressions in writing.

## OVERVIEW

Clarity, correctness, completeness, and conciseness are required in all technical writing. Clarity, however, is probably the most important because, when meaning remains obscure, nothing is achieved.

Clarity in writing results, in part, from correct grammar and spelling; from correct use of words, numbers, and capital letters; and from carefully developed sentences. As explained in this unit, it also results from using meaningful transitions, writing paragraphs correctly, using lead-in sentences, using concrete words, using action verbs, omitting negative expressions, and omitting substandard expressions.

# MEANINGFUL TRANSITIONS

In technical writing, a transition is a change from one subject to another. A transition is achieved through using words, phrases, and clauses that relate a new topic to one previously discussed. Because transitional words establish coherence and unity, they are important aids to clarity. Some words and phrases are used only for transition; these are discussed first. Lead-in sentences that are both transitional and introductory are discussed later in this unit. Headings as transitional aids are discussed in Unit 7.

Transitional words are like road signs; they indicate what direction the reader is taking. They tell whether a new sentence continues in the same direction as the previous thought or whether a comparison, a contrast, a conditional idea, a time relationship, or a result may be anticipated. Transitions are not needed for every sentence  or even for every paragraph. The writer who knows when to use them, however, helps the reader follow information logically and clearly.

Read the following two paragraphs. The first paragraph shows how each sentence becomes an independent element. It also shows how continuity of thought is lost when a necessary transitional word is omitted. The second paragraph shows the effectiveness of transitional words if they relate one idea to another meaningfully.

**EXAMPLES**

*No transitional words:* All fluids have some viscosity; viscosity may vary widely from one fluid to another. Viscosity is the property of a substance that makes that substance resist flow. High viscosity causes fluids to flow slowly; low viscosity causes fluids to flow freely. Alcohol flows more freely than glycerin and is less viscous than glycerin.

*Transitional words:* All fluids have some viscosity; *however,* viscosity may vary widely from one fluid to another. *Because* viscosity is a property that makes a substance resist flow, high viscosity causes fluids to flow slowly; low viscosity causes fluids to flow freely. *For example,* alcohol flows more freely than glycerin and is, *therefore,* less viscous than glycerin.

Some words and phrases that may be used for meaningful transitions are listed in the following examples. The general headings tell the direction in which a sentence will move in relation to the preceding sentence.

**EXAMPLES**

To continue ideas, use
  also
  and
  in addition
  moreover

To contrast ideas, use
  but
  conversely
  however
  in contrast
  on the contrary
  on the one hand ... on the other
    hand
  still

To give results of a previous idea, use
  as a result
  consequently
  for this reason
  hence
  therefore
  thus

To introduce a condition, use
  although
  if
  nevertheless
  under these circumstances

To show comparison, use
  besides
  furthermore
  in the same way
  likewise
  similarly

To introduce reasons, use
  because
  since

To show time sequence, use
  after
  a year later
  at present
  before this
  formerly
  later
  meanwhile
  now
  occasionally
  subsequently
  then
  when

To illustrate, use
  for example
  for instance

To sum up ideas, use
  in conclusion
  in summary

In addition to the words listed in the example, a noun repeated from a foregoing sentence may be a transitional word because it establishes unity between the two sentences. See the next example.

<table>
<tr><td>

---
**EXAMPLE**
---

</td><td>

**The self-inductance of a coil of wire opposes any changes in the flow of current.
This self-inductance may be expressed mathematically.**

</td></tr>
</table>

One other effective transition technique, especially in technical communications, is the use of numbered items. The following example illustrates the technique.

---
**EXAMPLE**
---

**Three invariable dilutions to the original plasms are (1) the addition of 1/11 volume
of thrombin, (2) the addition of 3/4 volume of incubation mixture, and (3) the
addition of 1/5 volume of fibrinogen.**

# PARAGRAPHING

A paragraph is a group of closely related sentences arranged in a way that permits a central idea to be defined, developed, and clarified. Skillfully used paragraphs made up of well-planned sentences are aids to clarity for two reasons: First, they permit a writer to follow a systematic development of the subject. Second, they prevent the reader from being subjected to an uninterrupted mass of words that creates confusion and monotony.

## Paragraph Development

Because technical writing is informative, every paragraph, except one beginning with a lead-in sentence (see the "Lead-in Sentences" section), must begin with a topic sentence. When a lead-in sentence is also used, the topic sentence immediately follows it.

A *topic sentence* introduces the subject to be developed within the paragraph. The rest of the paragraph accomplishes only one objective: It gives all details needed to develop this subject. The limits of a paragraph, then, are clearly defined; and decisions about beginning a new paragraph are based upon the question "Has the subject introduced by the topic sentence been completely developed?" If so, a new paragraph is begun.

The following paragraph begins with a topic sentence; then, explanation and examples develop the central idea.

**EXAMPLE**

Every fraction can be written in decimal form. All decimal forms, however, are not exact equivalents of the fractions for which they are substituted. Many fractions have exact decimal equivalents. For example, 1/4 is equal to 0.25; 3/5, equal to 6/10, is also equal to 0.60; and 1/20 is equal to 0.05. In contrast, other common fractions can be expressed only in approximate decimal numbers. For example, 1/3 can be approximately expressed as 0.3, 0.33, 0.333, or 0.3333. Each successive approximation is closer to the actual value of 1/3, but an exact decimal can never be obtained.

## Paragraph Length

Paragraph length, although determined by the subject being developed, should vary. A series of very long paragraphs presents detailed information too rapidly and will subject the reader to continuously long periods of concentration. A series of very short paragraphs forces the reader to change subjects too frequently. Therefore, paragraphs should range from about five to eighteen lines; ten to twelve lines is average length. If a paragraph is less than five lines, the topic sentence has probably been omitted or not fully developed. If a paragraph extends beyond eighteen lines, it may contain unrelated material or unnecessary words and phrases. An unusually long paragraph that cannot be shortened is more easily read if lists, graphic aids, or quotations are used to create open space within the solid block of words.

# LEAD-IN SENTENCES

Lead-in sentences, primarily introductory, improve clarity in communications because they introduce each major part of a report. Specifically, effective technical writers use a lead-in sentence to introduce the purpose of the report. They use another one for each major topic, as illustrated in the sample reports in Section III.

A lead-in sentence is not a substitute for a topic sentence but may precede one at the beginning of a paragraph, as illustrated in the following example.

**EXAMPLE**

All matter is divided into three types: (1) gaseous, (2) liquid, and (3) solid.

This sentence leads the reader toward a discussion of three separate topics each of which would probably be defined, discussed, and illustrated in a separate paragraph. Each paragraph, including the first one, would have its own topic sentence. If however, the writer intends only to define gaseous, liquid, and solid, the sentence given in the example is a topic sentence and one paragraph is written.

Lead-in sentences sometimes refer to previous information and, at the same time, introduce new topics. Under these circumstances, lead-in sentences are not only introductory but also transitional.

---

**EXAMPLE**

**After the chemicals have been mixed and cooled, the prints are exposed and processed.**

---

The sentence in this example tells the reader that the discussion about preparing chemicals is finished and that a new subject is being introduced. A writer who wants to keep the reader informed must decide whether the lead-in sentences should be both transitional and introductory or just introductory.

# CONCRETE WORDS

No one technique contributes more to clarity in technical writing than the use of concrete nouns and verbs. Concrete words refer to a specific, not a general, class. However, a clear-cut guideline for determining concrete words is difficult. The terms *specific, concrete, general,* and *abstract* are relative when applied to words because words vary in degree of concreteness and generality. For example, *animal* is a general word; *horse* is also general, though not so general as *animal.* When a general class of horse such as *draft horse* is named, the noun becomes more concrete, though not so concrete as a specific kind of draft horse such as *Clydesdale.* The most specific term gives a name such as *Nell* to one Clydesdale horse.

If many people in a group are asked to write a word that means tool, a variety of responses may be expected. *Wrench, hammer, screwdriver,* and *pliers* are some likely answers. Some people may also name garden or kitchen tools. The term *tool* is so general that people respond differently to it. If the word hammer is presented to the group, each person will at least think of a pounding tool. *Clawhammer* will bring a more unified response than *hammer* did because it is more concrete than the word *hammer.* A test of this kind indicates two things:

1. Words are only symbols that have no meaning until someone gives them meaning.
2. A reader's correct interpretation of a report is determined by the frequency with which the writer uses available concrete words.

Concrete nouns and verbs not only clarify meaning but also add vividness to a report. For example, readers are more interested in *gold, silver, copper,* and *aluminum* than they are in *metals.* They respond more to *welding* and *soldering* than they do to *joining together.* Consider the differences in tone, meaning, and interest between the general and concrete words listed in the following examples.

**EXAMPLES**

| General | Concrete |
| --- | --- |
| appliance | toaster, mixer, can opener |
| building | shop, warehouse, storeroom |
| do | perform, carry out, finish |
| doctor | heart surgeon, neurologist, oculist |
| figure (noun) | graph, chart, table, map |
| get | grasp, seize, hold, receive |
| hand tools | wrench, pliers, hammer, chisel |
| join | wire, solder, weld |
| lawn mower | rotary power mower |
| make a hole | drill or bore a hole |
| make a part | build or construct a part |
| mechanical device | gear, link, chain, cam |
| mineral | asbestos, clay, mica |

Note that a specific kind of screwdriver, plier, hammer, or other tool is preferable to the more general classification.

**EXAMPLES**

—*Offset screwdriver* is preferable to *screwdriver.*
—*Long-nose pliers* is preferable to *pliers.*

Sometimes, a special name or size of tool increases clarity.

**EXAMPLES**

—a Phillips screwdriver
—a 1/2-inch drill
—one 24-inch bolt-and-rod cutter

The examples show that modifiers such as adjectives help make words more concrete.

# ACTION VERBS

Verbs that show action are aids to clarity because they give force and motion to ideas. The verb *to be* (am, is, are, was, were, be, being, and been) does not express action; it expresses only existence. Although an important word, both as a helping and a main verb, *to be* and its various forms can weaken the text. As explained earlier, writers can usually avoid using *there is* and *there are*. They can also consider the following guidelines.

- Substitute a form of the verb *measure* for *is, are, was,* or *were* in sentences concerning size.

**EXAMPLE**

*Original:* **The shaft was 0.25 of an inch long.**

*Revision:* **The shaft measured 0.25 of an inch.**

- Substitute other words for *is* and *are* in sentences introducing lists of items.

**EXAMPLE**

*Original:* **The three operations needed to complete the process are as follows:**

*Revision:* **The process involves three separate operations.**

- Check to see whether the best word has been used as the subject of the sentence.

**EXAMPLE**

*Original:* **Two resistors and two power supplies are in the circuit.**

*Revision:* **The circuit has two resistors and two power supplies.**

- Avoid using adjective clauses beginning with *that is* or *that are* when adjectives, participles, or prepositional phrases can be substituted.

**EXAMPLE**

*Original:* **The tubes *that are defective* must be discarded. (Change the dependent clause to an adjective.)**

*Revision:* **The *defective* tubes must be discarded.**

*Original:* **The formula *that was used to determine power* was incorrect. (Change the dependent clause to a participial phrase.)**

*Revision:* **The formula *used to determine power* was incorrect.**

*Original:* **The mercury *that is in the tube* expands when heat is applied. (Change the dependent clause to a prepositional phrase.)**

*Revision:* **The mercury *in the tube* expands when heat is applied.**

# NEGATIVE EXPRESSIONS

Negative expressions in formal reports not only include words that say "no" but also include any expressions that cause a reader to question the accuracy of information. When justifications, limitations, or incomplete data interrupt statements of facts, the facts become less impressive that the explanatory discussions. Consider the following example.

**EXAMPLE**

**The motors seem to be operating at 75-percent efficiency; however, data is incomplete.**

Reading this sentence, a person may begin to question how or why the data is incomplete and whether the information is valid. The primary information loses effectiveness. The statement can be written in a positive style, as shown next.

**EXAMPLE**

**Current data indicates that the motors are operating at 75-percent efficiency.**

The reader now realizes that the data may be incomplete but is likely to concentrate upon the efficiency of the motors.

Even a positive statement of qualification is rarely used in the body of a report. When many topics require limitations similar to that in the example, general statements of limitation are made in the report introduction, as shown in the basic outline of Unit 6. Then, the available facts can be presented without interruption.

In report writing, verbs such as *seem, believe, may, might,* and *could* and modifiers such as *possibly, probably, perhaps, almost, approximately, nearly,* and *unfortunately* often suggest doubt. They should, therefore, be used only after a writer has evaluated their implications.

**EXAMPLE**

**The air hose possibly can be repaired.**

*Possibly* has prevented any positive information. The sentence, therefore, has little value.

# SUBSTANDARD ENGLISH

Substandard English is unacceptable in formal communications because it distracts the reader and is easily misinterpreted, thus obstructing clarity. Words and phrases classed as substandard are slang expressions, colloquialisms, clichés, and some technical jargon. Some specific expressions in each category are listed in the next example.

**EXAMPLE**

**Clichés**

a good many

all in all

as a matter of fact

at one time or another

best laid plans

few and far between

first and foremost

in this day and age

it goes without saying

last but not least

now and then

**Colloquialisms**

ace in the hole

airing, as in *airing the problem*

all-out, as in *an all-out effort*

A–1, meaning outstanding

flabbergasted

fired-up, as in "The transmitter was fired up."

first off

take a try at

tote, meaning to carry

**Slang**

beef up, meaning to add strength

broken up, meaning divided into separate parts

clear sailing, meaning easy to accomplish

fouled up, meaning ruined as a result of incompetence

know-how

tough, as in *tough decision*

yes-man

**Technical Jargon**

amp for ampere

bio for biology

calc for calculus

ceiling, meaning limit

econ for economics

inorgan for inorganic chemistry

lab for laboratory

trig for trigonometry

trigger, meaning to start or begin

tech for technology or technological

*Slang* includes any word that has been coined by a few people and temporarily adopted by others. A particular slang word often endures for only a short time. Then, it is replaced by another slang word. A *colloquialism* is an expression characteristic of one locality or group of people. It has no universal meaning. A *cliché* is a wordy phrase that has been repeated so frequently that any meaning it once had is destroyed by the monotonous tone it creates.

One other kind of substandard English is known as *technical jargon.* Technical jargon is a word, frequently a shortened technical word, that is understood by people who work together but is meaningless to others.

Review the preceding example. Remember that none of the expressions listed, or similar expressions, are acceptable in technical writing.

# SUMMARY

1. Clarity is possibly the most important quality in any technical communication. Unless a reader understands written text, its accuracy, completeness, and conciseness have little value.

2. Coherence — that is, keeping ideas tied closely to one another — contributes to clarity. Transitional words, phrases, and clauses and lead-in sentences help develop coherence.

3. Short paragraphs logically developed from a topic sentence allow information to develop logically and, therefore, help create clarity.

4. Using concrete words and active verbs to express information contributes not only to clarity but also to reader interest.

5. Any distraction such as negative or qualifying statements that diminish unity and coherence lessen clarity.

6. Substandard English, including clichés, colloquialisms, and technical jargon, distracts a reader from the text and can thus adversely affect clarity.

UNIT **30**

# SUGGESTED ACTIVITIES

A. Select a paragraph from a textbook or supplementary reading. Analyze the paragraph in relation to the following questions.
   1. Does the paragraph begin with a topic sentence?
   2. Does the paragraph fully develop one central idea?
   3. Are transitional words used? Have any been omitted that should be included?

   4. Are any substandard English expressions used?

   5. Are any ideas interrupted by words that suggest doubt?

   6. Can any words or sentences be changed to make the paragraph more easily understood?

   7. Do you think the paragraph is well written? Give specific reasons for your answer.

B. From information learned in other classes, select an idea that can be developed in one paragraph. Write a topic sentence; then, develop the paragraph. Write objectively. Use standard English, and use transitional words when they are needed for clarity. Choose a simple subject. For example, a specific resistor, spark plug, or other small part may be defined, explained, and illustrated. A term such as *fraction* or *mixed number* in mathematics can be discussed; or a word such as *valence, energy,* or *matter* may be clarified. Any of these suggested topics can probably be extended beyond a paragraph. The important objective in this activity, however, is to develop one topic sentence.

C. Rewrite the following sentences by using standard English and eliminating substandard expressions.

   1. It goes without saying that this transmitter is the best one for the money.

   2. The supervisor said that the inventory must be made, but that's easier said than done.

   3. The last but not the least problem is determining the output voltage of a three-phase generator.

   4. The accidents resulting from explosions in the shop are few and far between — fewer than two a year.

   5. An automobile power train is broken up into three major parts.

   6. Technicians should occasionally brush up on new techniques relating to their work.

   7. After the engine was fired up, the power output was determined.

   8. Because the circuit was fouled up, the circuit breaker tripped.

D. Using an objective, standard English style, rewrite the following paragraph. Observe all guidelines for correct grammar, sentence structure, use of numbers, capital letters, abbreviations, and symbols. Make the meaning clear and specific. To be specific, assume that all measurements used in the original paragraph are exact. You may rearrange sentences if necessary to permit organized presentation of information.

   The object that I am going to describe is a thermometer which is used for measuring temperature is made up of a clear glass tube about six inches long and is about 1/4 in. in diameter. (See fig. 1.) The bottom of the tube opens up into a bulb that is 1/2 inches in diameter. This tube is mounted on a piece of metal, probably aluminum, painted white, by 2 staples one at the top, maybe about 1/2 inch down and one at the top of the bulb. The plate is 7 1/2 inches by 2 inches. On either side of the tube, painted on the metal border are black Nos. which are used to designate the temperature, -60 being the bottom number and 130 being the top number.

SECTION **VI**

# SUGGESTED SUPPLEMENTARY READING

## Grammar Handbooks and Grammar Reference Books

Copperud, Roy H. *American Usage: The Consensus*. New York: Van Nostrand Reinhold Company, 1970. The text of this book discusses varying authoritative points of view concerning rules of grammar and the use of words. It helps students develop judgment and insight concerning the effective use of language.

*Guidelines for Nonsexist Use of Language in NCTE Publications*. Urbana, Ill.: National Council of Teachers of English, Stock No. 19719, 1975.

Perrin, Porter G. *Reference Handbook of Grammar & Usage* (derived from *Writer's Guide and Index to English* by Porter G. Perrin). New York: William Morrow & Company, Inc., 1972. This book contains excellent information concerning the correct use of grammar but has no table of contents or index. The subject matter concerning grammar and usage, however, is listed alphabetically. Each topic, such as "Principal Parts of Verbs," is discussed independently.

Rosen, Leonard. *The Everyday English Handbook*. Garden City, N.Y.: Doubleday & Company, Inc., 1985. A detailed study of grammar, sentence structure, and paragraphing.

Williams, Joseph M. *Style*. Glenview, Ill.: Scott, Foresman and Company, 1985. This small book is interesting, informative, and practical. Written to show that anyone can write clearly, it uses several interesting quotations related to the importance of language and its effective use. Its content concerns clarity, coherence, concision (conciseness), and general discussions of sentence structure, style and usage, and style and punctuation.

Wilson, Robert F., John M. Kierzek, and W. Walker Gibson. *The Macmillan Handbook of English*. New York: Macmillan Publishing Company, Inc., 1982. Important parts of this book for technical writers include "Using the Dictionary," pages 42 to 53; "Organizing Paragraphs in Sequences," pages 154 to 158; Chapter 9, "Grammar and Usage"; and Chapter 13: "Effective Use of Words and Phrases."

## General Word References

Flesch, Rudolf. *The ABC of Style*. New York: Harper and Row Publishers, 1980. In *The ABC of Style* an interesting evaluation of several misused, overused, and unnecessary words is given.

Fowler, H. S. *A Dictionary of Modern English Usage.* New York: Oxford University Press. This reference book has long been recognized as an excellent source of information concerning the proper use of English. The author incorporates both common sense and wit into his discussion about numerous words.

*Roget's Thesaurus.* New York: Thomas Y. Crowell Company (any recent edition).

*Webster's Collegiate Thesaurus.* Springfield, Mass.: G & C Merriam Company, 1985 (or any recent edition).

*Webster's Ninth New Collegiate Dictionary.* Springfield, Mass.: Merriam-Webster Inc., 1983 (or any recent edition).

## Other Reading for Section VI

Flesch, Rudolph, and A. H. Lass. *A New Guide to Better Writing.* New York: Warner Books, Inc., 1984. This book contains the following information useful to a technical writer: transitional techniques, pages 55 to 65; ideas for using words, pages 88 to 95; correct words, pages 233 to 257; and an easily understood distinction between plural and possessive nouns, pages 285 to 289.

Houp, Kenneth W., and Thomas E. Pearsal. *Reporting Technical Information.* New York: Macmillan Publishing Company, Inc., 1984. Common grammatical errors are illustrated and explained on pages 453 to 463.

Lannon, John M. *Technical Writing.* Boston: Little, Brown and Company, 1985. This book discusses efficient sentences on pages 42 to 57, exact words on pages 57 to 65, and sexist language on page 72.

Markel, Michael H. *Technical Writing: Situations and Strategies.* New York: St. Martin's Press, Inc., 1984. Choosing between active and passive voice is effectively illustrated on pages 90 and 91. A readability check for sentence structure is given on page 105. A detailed discussion of paragraphing is found on pages 110 to 119. Grammar and punctuation are discussed on pages 491 to 500.

Rathbone, Robert R. *Communicating Technical Information.* Reading, Mass.: Addison-Wesley Publishing Company, Inc., 1966. The following topics, "The Artful Dodge," pages 41 to 49, and "The Ubiquitous Noise," pages 51 to 62, are interesting and informative discussions of clarity, accuracy, and conciseness.

Roundy, Nancy, with David Mair. *Strategies for Technical Communication,* Part V: "Handbook of Style: Capitalization, Grammar, Punctuation, Spelling, Usage." Boston: Little, Brown and Company, 1985.

Swindle, Robert E. *The Concise Business Correspondence Style Guide,* Section I: "Getting Ready to Communicate." Englewood Cliffs, N.J.: Prentice-Hall, Inc., 1983.

# INDEX